INTEGRABLE
MODELS

World Scientific Lecture Notes in Physics, Vol. 30

INTEGRABLE MODELS

Ashok Das

University of Rochester
Rochester, New York

World Scientific
Singapore • New Jersey • London • Hong Kong

Published by

World Scientific Publishing Co. Pte. Ltd.
P O Box 128, Farrer Road, Singapore 9128

USA office: World Scientific Publishing Co., Inc.
687 Hartwell Street, Teaneck, NJ 07666, USA

UK office: World Scientific Publishing Co. Pte. Ltd.
73 Lynton Mead, Totteridge, London N20 8DH, England

INTEGRABLE MODELS

ISBN 9971-50-910-5
        9971-50-911-3 (pbk)

Printed in Singapore by JBW Printers & Binders Pte. Ltd.

To
My Brothers and Sisters,
Especially,
To the Memory of
Nilu.

PREFACE

The subject of integrable models - both classical and quantum - is fascinating. Decades of research in this area has led to mathematical developments which are quite beautiful and which unify various aspects of physical problems that appear to be disparate. These special developments, however, have not yet been widely appreciated. The primary reason for this appears to be the technical nature of the subject. The purpose of this book, therefore, is to present the basic ideas in a systematic manner so that people in different areas may share the excitement and find the methods useful in their respective areas of research.

This book grew out of a graduate course which I taught in the Fall of '88 in the Department of Physics and Astronomy at the University of Rochester. In fact, except for the last few chapters, the material in this book is exactly what was presented in the class. Consequently, derivations of various equations appear in explicit detail and the reader is free to concentrate only on the logical development of the ideas. Furthermore, most of the techniques are developed with specific examples in order to make ideas more transparent. The first eight chapters of the book develop various ideas of classical integrability with the example of the Korteweg-de Vries equation. The geometrical approach as well as the group theoretical approach to integrable models are explained with the Toda lattice as an example. Finally, the methods of zero curvature and quantum inverse scattering are discussed within the context of the nonlinear Schrödinger equation. The level of discussion has been carefully chosen so as to make the material accessible to graduate students.

All the models I have considered are defined on the real line. Lack of space did not permit me to discuss the periodic systems although I have tried to comment, wherever possible, on the generalization of various formulae to such cases. I have also not been able to discuss either the method of Hirota or the supersymmetric systems. Every chapter is supplemented with a list of references which I found useful in the preparation of the lecture notes. They are by no means meant to be a complete set.

There are several people whose help has been invaluable in the preparation of this manuscript. First of all, I would like to thank all the students in the class for providing a stimulating atmosphere and I would like to thank, in particular, Messrs. T. Blum, S. Borzillary and S. Ramaswamy for various forms of assistance during the preparation of the notes. I gratefully acknowledge the encouragement and support of Prof. P. Slattery. I have benefited enormously from discussions with Profs. J. Maharana, V. S. Mathur and C. Taylor and appreciate their comments and suggestions. I am truly indebted to Prof. S. Okubo who has painstakingly read through the manuscript and has generously clarified many ideas. It is also a pleasure to thank Prof. A. C. Melissinos for sharing with me his enthusiasm which has helped me through the long hours during the preparation of the manuscript. An Outstanding Junior Investigator Award from the U.S. Department of Energy during the period when this book was written is also gratefully acknowledged.

The credit for the get up of the book goes completely to Judy Mack who has gracefully typed the manuscript and has accommodated all the changes in a very short time. And finally, I would like to express my heartfelt gratitude to Ammani whose unending desire for perfection has been a constant source of inspiration to me.

                                        Ashok Das

CONTENTS

PREFACE   vii

1. THE KORTEWEG-DE VRIES EQUATION   1

   Hamiltonian systems   2
   Integrability of a Hamiltonian system   7
   KdV equation   11
   KdV as a Hamiltonian system   14
   References   22

2. PROPERTIES OF THE KDV SOLUTIONS   24

   Uniqueness of solutions   24
   Travelling waves   28
   Solitons   30
   Quantum Mechanics of the soliton solutions   34
   References   45

3. INTEGRABILITY OF THE KDV EQUATION   46

   Conserved quantities   46
   The Miura transformation   51
   Infinite number of conserved quantities   55
   Integrability of the KdV system   65
   Higher order equations of the hierarchy   71
   References   73

4. INITIAL VALUE PROBLEM FOR THE KDV EQUATION   74

   The Schrödinger equation   74
   Time evolution of the scattering parameters   80
        i) Bound states   80
       ii) Scattering states   82

Gel'fand-Levitan equation   85
Example   86
References   91

5.   **INVERSE SCATTERING THEORY**   92

Scattering in one dimension   92
Analytic behavior of scattering coefficients   96
Action-angle variables for KdV   105
References   116

6.   **THE LAX METHOD**   117

Origin of the Schrödinger equation   117
The Lax pair   121
Specialization to KdV   125
Alternate construction   129
Examples   131
Lenard's derivation of the KdV equation   135
References   139

7.   **MORE ON KDV**   140

The spectral parameter   140
Involution of the conserved quantities   150
KdV equation and the group SL(2,R)   153
References   161

8.   **MULTI-SOLITON SOLUTIONS**   162

Bäcklund transformations   162
Examples: Liouville equation   163
            Sine-Gordon equation   167
Theorem of permutability   172
Bäcklund transformation for the KdV equation   174

Soliton solutions   180
References   188

9. GEOMETRICAL APPROACH TO INTEGRABLE MODELS   189

Symplectic Geometry   189
Integrable models   196
Example: Geometrical approach to KdV   207
References   211

10. THE TODA LATTICE   213

The Toda equation   213
Dual Poisson bracket structure   215
Conserved quantities   219
The Nijenhuis tensor   221
The Lax equation   225
References   230

11. GROUP THEORETICAL APPROACH TO THE TODA LATTICE   232

Review of Lie algebra   232
Group structure of the Toda equations   240
The Lax pair   244
Integrability of the Toda system   247
References   252

12. ZAKHAROV-SHABAT FORMULATION   253

The first order formulation   253
The nonlinear equations: 259
     i)   KdV   259
    ii)   MKdV   260
   iii)   Nonlinear Schrödinger equation (NSE)   262
    iv)   Sine-Gordon equation   263
     v)   Sinh-Gordon equation   264

NSE as a Hamiltonian system   265
Time development of the scattering data   266
References   277

13. THE ZERO CURVATURE METHOD   278

The zero curvature condition   278
The transition matrix   281
Time evolution of the transition matrix   285
Fundamental Poisson bracket relations   288
Action-angle variables   292
References   302

14. QUANTUM INTEGRABILITY   303

The Bethe Ansatz   303
Quantum Inverse Scattering   309
Yang-Baxter Equation and Quantum Groups   320
References   325

APPENDIX   327

INDEX   339

INTEGRABLE
MODELS

CHAPTER 1

THE KORTEWEG-DE VRIES EQUATION

In the course of these lectures, we will discuss various nonlinear models which are integrable. These describe systems of nonlinear differential equations which can be solved exactly. Most of these models would be continuum models in one space and one time dimension although we will also study the Toda lattice which consists of only a finite number of degrees of freedom. All of these models are quite interesting both from the mathematics as well as the physics points of view. Mathematically, we encounter new concepts such as the infinite dimensional Lie algebras and their representation. From the physics point of view, these models describe physical phenomena in such diverse areas as nonlinear optics, hydrodynamics, condensed matter, plasma physics, high energy physics and so on. It is, therefore, not surprising that physicists and mathematicians alike have found this area of research fascinating for over two decades now. It is, however, surprising that a lot still remains to be understood in this area and active research in this field is presently being pursued by several distinct groups all over the world.

The study of integrable models has been approached from various points of view. In order to fix ideas clearly, we will describe these different approaches with the same example of the integrable system, namely, the Korteweg-de Vries equation. In the later chapters, we will study several other models which are also soluble. In this chapter, we will review the various necessary concepts and introduce the Korteweg-de Vries (KdV) equation. All the models we will study are Hamiltonian in nature and their solubility is intimately associated with the existence of soliton solutions. So, we begin by reviewing what a Hamiltonian system is, what the conditions of integrability are and how soliton solutions may be appropriate for an integrable system.

Hamiltonian Systems:

A Hamiltonian system is a system whose evolution can be described by a set of Hamilton's equations.

Frequently, in a mechanical system, we are given a set of canonical coordinates and momenta q_i, p_i with i = 1,....N, and a Hamiltonian H such that the evolution equations take the form

$$\dot{q}_i = \{q_i, H\}$$

$$\dot{p}_i = \{p_i, H\} \qquad (1.1)$$

Here the curly brackets represent Poisson brackets and for simplicity we have chosen to work with a system with a finite number of degrees of freedom. It is clear, therefore, that a Hamiltonian system consists of dynamical variables (q_i, p_i), a Hamiltonian and the fundamental Poisson bracket relations which in this example happen to be canonical, namely,

$$\{q_i, q_j\} = 0 = \{p_i, p_j\}$$

$$\{q_i, p_j\} = \delta_{ij} \qquad (1.2)$$

so that

$$\dot{q}_i = \{q_i, H\} = \frac{\partial H}{\partial p_j} \{q_i, p_j\} = \frac{\partial H}{\partial p_i}$$

$$\dot{p}_i = \{p_i, H\} = \frac{\partial H}{\partial q_j} \{p_i, q_j\} = - \frac{\partial H}{\partial q_i}$$

(1.3)

The dynamical variables (q_i, p_i) span the phase space. Namely, they are the coordinates of the phase space which is 2N dimensional. Consequently, a split of the coordinates into canonical coordinates and momenta is noncovariant and artificial. One could rectify this by combining the coordinates and momenta into a set of 2N generalized coordinates.

$$y^\mu = (q_i, p_i)$$

$$i = 1, 2 \ldots N \quad (1.4)$$

$$\mu = 1, 2 \ldots 2N$$

such that

$$y^i = q_i$$

$$y^{N+i} = p_i$$

(1.5)

The fundamental Poisson brackets (canonical ones) then take the form

$$\{y^\mu, y^\nu\} = \epsilon^{\mu\nu} \tag{1.6}$$

where

$$\epsilon^{\mu\nu} = \begin{pmatrix} 0 & I \\ -I & 0 \end{pmatrix} \tag{1.7}$$

is the constant antisymmetric 2N×2N matrix written in terms of N×N blocks. This implies that

$$\{A(y), B(y)\} = \partial_\mu A \{y^\mu, y^\nu\} \partial_\nu B$$

$$= \partial_\mu A \, \epsilon^{\mu\nu} \partial_\nu B = \epsilon^{\mu\nu} \partial_\mu A \, \partial_\nu B \tag{1.8}$$

where we have defined

$$\partial_\mu = \frac{\partial}{\partial y^\mu} \tag{1.9}$$

Hamilton's equations can now be written covariantly as (compare with Eq. (1.3))

$$\dot{y}^\mu = \{y^\mu, H\} = \epsilon^{\mu\nu} \partial_\nu H \tag{1.10}$$

The advantage of the covariant method is that there are many physical systems that are better described in terms of noncanonical coordinates. The most familiar example of noncanonical coordinates is in systems described in terms of constrained variables. In this case, the covariant formalism gives a similar description of Hamilton's equations as in Eqs. (1.6) and (1.10) , namely,

$$\dot{y}^\mu = \{y^\mu, H\} = f^{\mu\nu}(y)\partial_\nu H \qquad (1.11)$$

with

$$\{y^\mu, y^\nu\} = f^{\mu\nu}(y) \qquad (1.12)$$

That is, in such a case, the fundamental Poisson brackets become coordinate dependent. (This is familiar from the studies of constrained systems where the canonical Poisson brackets are eventually replaced by Dirac brackets which are in general coordinate dependent.) Similarly, in this case the analogous formula for Eq. (1.8) becomes

$$\{A(y), B(y)\} = \partial_\mu A(y) f^{\mu\nu}(y)\partial_\nu B(y) \qquad (1.13)$$

Let us note here that $f^{\mu\nu}(y)$ must satisfy various properties in order to describe the Poisson bracket of Eq. (1.12). For example,

$$\{y^\mu, y^\nu\} = -\{y^\nu, y^\mu\}$$

and, therefore   (1.14)

$$f^{\mu\nu}(y) = -f^{\nu\mu}(y)$$

Furthermore, for y^μ to form a basis (linear independence) of the phase space, $f^{\mu\nu}(y)$ must be nonsingular as a matrix. Let us denote the inverse by $f_{\mu\nu}(y)$ such that

$$f_{\mu\lambda}f^{\lambda\nu} = \delta^{\nu}_{\mu} = f^{\nu\lambda}f_{\lambda\mu} \qquad (1.15)$$

Finally, the definition of a Poisson bracket must satisfy the Jacobi identity. Namely, in terms of the fundamental variables

$$\{y^{\mu},\{y^{\nu},y^{\lambda}\}\} + \{y^{\lambda},\{y^{\mu},y^{\nu}\}\} + \{y^{\nu},\{y^{\lambda},y^{\mu}\}\} = 0 \qquad (1.16)$$

This condition translates to a simple condition on the $f_{\mu\nu}(y)$'s, namely,

$$\partial_{\mu}f_{\nu\lambda}(y) + \partial_{\lambda}f_{\mu\nu}(y) + \partial_{\nu}f_{\lambda\mu}(y) = 0 \qquad (1.17)$$

In other words, the inverse of the Poisson bracket structure must satisfy the Bianchi identity.

A manifold with a preferred second rank antisymmetric tensor which satisfies the above conditions is called a symplectic manifold. Clearly the geometry of the dynamics of a Hamiltonian system is automatically symplectic. The main differences of a symplectic geometry from a Riemannian geometry are that we do not have a symmetric metric for the symplectic geometry and secondly, in a symplectic manifold the symmetry group of the tangent space is the symplectic group whereas the orthogonal group is the symmetry group of the tanget space for a Riemannian manifold. The tensors $f^{\mu\nu}(y)$ and $f_{\mu\nu}(y)$ can, however, be thought of as the contravariant and covariant components of the metric tensor

in a symplectic manifold.   They are known as the symplectic
metric and can be used to raise or lower indices in a
symplectic manifold analogous to the Riemannian metric in a
Riemannian manifold.

For completeness let us also record here that if instead
of a system with a finite number of degrees of freedom, we
are dealing with a continuum system, all the formulae go
through if we replace partial derivatives by functional
derivatives. For example, Eqs. (1.12) and (1.11) take the
respective forms

$$\{u(x),u(y)\} = f(x,y) \tag{1.18}$$

$$\dot{u}(x) = \{u(x),H\} = \int dy\, f(x,y)\, \frac{\delta H}{\delta u(y)} \tag{1.19}$$

Furthermore, the Poisson bracket of two functionals $A[u]$ and
$B[u]$ is obtained to be (compare with Eq. (1.13))

$$\{A[u],B[u]\} = \int dx\, dy\, \frac{\delta A}{\delta u(x)}\, f(x,y)\, \frac{\delta B}{\delta u(y)} \tag{1.20}$$

Integrability of a Hamiltonian System:

Given a Hamiltonian system, we can ask when it is that
it would be integrable.   Or, more precisely, what is the
criterion of integrability?   The study of this question
dates back to Liouville who concluded the following.

A Hamiltonian system whose phase space is 2N dimensional
is integrable by the method of quadratures if and only if

there exist exactly N, functionally independent conserved quantities which are in involution.   (That is, the Poisson brackets of these conserved quantities with one another vanish.)

Let us not go into the proof of Liouville's theorem which exists in many texts on classical mechanics. Rather, let us understand its implications.

Let us assume that the system is describable by the canonical coordinates and momenta (q_i, p_i), $i = 1,2,....N$. Let K_i, $i = 1,2,....N$, represent the N functionally independent conserved quantities.   Clearly, the Hamiltonian must be a linear combination of these.   Furthermore, we assume that

$$\{K_i, K_j\} = 0 \qquad\qquad i,j = 1,2....N \qquad (1.21)$$

This implies that we can consider a canonical transformation

$$K_i = K_i(q_j, p_j) = P_i \qquad\qquad\qquad (1.22)$$

where we can think of the K_i's as the new momenta. These are also known as the action variables of the theory. Secondly, we also realize that we can, in principle, obtain the canonically conjugate variables

$$\theta_i = \theta_i(q_j, p_j) \qquad\qquad\qquad (1.23)$$

such that

$$\{\theta_i, \theta_j\} = 0 = \{P_i, P_j\} \tag{1.24}$$

$$\{\theta_i, P_j\} = \{\theta_i, K_j\} = \delta_{ij} \qquad i,j = 1,2 \ldots N \tag{1.25}$$

The coordinates θ_i are also known as the angle variables. Since

$$H = H(P_i)$$

Hamilton's equations (Eq. (1.3)) take the form

$$\dot{P}_i = \{P_i, H(P_j)\} = 0 \tag{1.26}$$

$$\dot{\theta}_i = \{\theta_i, H(P_j)\} = f_i(P_j) \tag{1.27}$$

Eq. (1.26) is simply a statement of the fact that P_i's are conserved. Since P_i's are conserved, for given values of these quantities, the equation for the angle variables, Eq. (1.27), takes the form

$$\dot{\theta}_i = f_i = \text{constant} \tag{1.28}$$

This can be readily integrated to give

$$\theta_i = f_i t + \alpha_i \tag{1.29}$$

where α_i's are integration constants which can be fixed from

the initial conditions. Thus we see that the motion or the
evolution can be completely fixed, in principle, in terms of
the action-angle variables (P_i, θ_i).   We say "in principle"
because it may be impossible in practice to determine the
canonical transformation

$$\theta_i = \theta_i(q_j, p_j) \tag{1.30}$$

and subsequently to invert the solution to obtain the
evolution as

$$q_i = q_i(f_j, \alpha_j, t)$$

$$p_i = p_i(f_j, \alpha_j, t) \tag{1.31}$$

Liouville's theorem only   asserts the existence of
solutions. The actual construction then involves various
ingenious procedures.

Although we discussed Liouville's theorem in the case of
a finite number of degrees of freedom, the same goes through
for a system with an infinite number of degrees of freedom
with a little modification. But it is clear that a system
with an infinite number of degrees of freedom, i.e., a
continuum system must have an infinite number of conserved
quantities. Consequently any solution of such a system must
be infinitely restricted.   A soliton is precisely such a
solution. Namely, it is a localized wave which retains its
shape (nondispersive) even after collisions. Intuitively,
it is clear that for this to happen, there must be an
infinite number of conservation laws.   Consequently,
solitons are implied by the existence of an infinite number

of conserved quantities and vice versa. Therefore, integrable models and solitons are often used interchangeably.

KdV Equation:

The KdV equation or the Korteweg-de Vries equation was formulated to explain the solitary water waves observed by J. Scott Russell in the Edinburgh Glasgow canal. It is a nonlinear equation in one space and one time dimension which as we will see shortly possesses soliton solutions. But at the time of its formulation, nothing was known about the integrability of the equation. The equation has the form

$$\frac{\partial u}{\partial t} + au \frac{\partial u}{\partial x} + b \frac{\partial^3 u}{\partial x^3} = 0 \qquad (1.32)$$

where $u = u(x,t)$ is the dynamical variable and "a" and "b" are arbitrary constants. In the case of the water waves, for example, $u(x,t)$ can be thought of as the height of the wave above the water surface. Furthermore, let us note that although the KdV equation involves two arbitrary constants, they can be eliminated by rescaling the variables. For example, a scaling

$$x \rightarrow b^{1/3} x \qquad (1.33)$$

takes the equation to the form

$$\frac{\partial u}{\partial t} + ab^{-1/3} u \frac{\partial u}{\partial x} + \frac{\partial^3 u}{\partial x^3} = 0 \qquad (1.34)$$

Next, if we let

$$t \rightarrow -t$$

the equation becomes

$$\frac{\partial u}{\partial t} = ab^{-1/3} u \frac{\partial u}{\partial x} + \frac{\partial^3 u}{\partial x^3} \qquad (1.35)$$

Finally, if we let

$$u \rightarrow a^{-1} b^{1/3} u \qquad (1.36)$$

we obtain

$$\frac{\partial u}{\partial t} = u \frac{\partial u}{\partial x} + \frac{\partial^3 u}{\partial x^3} \qquad (1.37)$$

In fact it is this form of the equation which we will refer to as the KdV equation. But let us note that any other form of the equation in the literature can be obtained through a simple transformation of the variables.

This also brings us to the question of symmetries of the KdV equation. Symmetries are very often helpful in finding solutions as well as classifying them. Let us note that the KdV equation does possess several symmetries. Namely,

i) $t \rightarrow t + c_1$ $\qquad\qquad\qquad\qquad$ (1.38) where c_1 is a constant, leaves the equation invariant.

ii) $x \to x + c_2$ (1.39)

where c_2 is a constant is also a symmetry of the KdV equation.

iii) $x \to cx$, $t \to c^3 t$, $u \to c^{-2}u$ (1.40)

leads to

$$\frac{\partial u}{\partial t} - u\frac{\partial u}{\partial x} - \frac{\partial^3 u}{\partial x^3}$$

$$\to c^{-5}\frac{\partial u}{\partial t} - c^{-5}u\frac{\partial u}{\partial x} - c^{-5}\frac{\partial^3 u}{\partial x^3}$$

$$= c^{-5}\left(\frac{\partial u}{\partial t} - u\frac{\partial u}{\partial x} - \frac{\partial^3 u}{\partial x^3}\right)$$ (1.41)

Therefore, the transformation in Eq. (1.40) is a symmetry of the KdV equation. In fact, this is the scaling symmetry which is quite useful in classifying the various conserved quantities of the system.

iv) $x \to x + vt$, $t \to t$, $u \to u + v$ (1.42)

where v is a constant, is also a symmetry. To see this, note that under this transformation

$$\frac{\partial u}{\partial t} - u\frac{\partial u}{\partial x} - \frac{\partial^3 u}{\partial x^3}$$

$$\rightarrow \frac{\partial u}{\partial t} + v\,\frac{\partial u}{\partial x} - (u+v)\,\frac{\partial u}{\partial x} - \frac{\partial^3 u}{\partial x^3}$$

$$= \frac{\partial u}{\partial t} - u\,\frac{\partial u}{\partial x} - \frac{\partial^3 u}{\partial x^3} \qquad\qquad (1.43)$$

This can be thought of as the Galilean invariance of the equation. Note, however, that the KdV equation cannot be Lorentz covariant since time and space are treated separately. (linear in $\frac{\partial}{\partial t}$ whereas higher order in $\frac{\partial}{\partial x}$.)

KdV as a Hamiltonian System:

Let us examine the KdV equation in Eq. (1.37), namely,

$$\frac{\partial u}{\partial t} = u\,\frac{\partial u}{\partial x} + \frac{\partial^3 u}{\partial x^3}$$

The fundamental variables, $u(x,t)$, in this case are continuous functions of x and t. The $u(x,t)$'s can be thought of as the generalized coordinates of the phase space (equivalent to the y^μ's of the finite dimensional case). The evolution equation is already first order in time - a feature inherent in the Hamilton's equations, for example, in Eqs. (1.11) and (1.19). To show that it is indeed a hamiltonian system, we have to find a fundamental Poisson bracket relation and a Hamiltonian which would give the KdV equation as the Hamilton's equation. That is, we would like to write

$$\frac{\partial u}{\partial t} = u \frac{\partial u}{\partial x} + \frac{\partial^3 u}{\partial x^3}$$

as   (1.44)

$$\frac{\partial u}{\partial t} = \{u(x,t),H\}$$

To be able to guess a Poisson bracket structure and a Hamiltonian for this system, let us rewrite equation (1.37) as

$$\frac{\partial u}{\partial t} = \frac{\partial}{\partial x} \left(\frac{1}{2} u^2\right) + \frac{\partial}{\partial x} \left(\frac{\partial^2 u}{\partial x^2}\right)$$

$$= \frac{\partial}{\partial x} \left(\frac{1}{2} u^2 + \frac{\partial^2 u}{\partial x^2}\right)$$ (1.45)

Note here, that if we choose as a Hamiltonian

$$H[u] = \int_{-\infty}^{\infty} dx \left(\frac{1}{3!} u^3 - \frac{1}{2} \left(\frac{\partial u}{\partial x}\right)^2\right)$$ (1.46)

then

$$\frac{\delta H}{\delta u(x)} = \frac{1}{2} u^2(x) + \frac{\partial^2 u(x)}{\partial x^2}$$ (1.47)

[For those unfamiliar with functional derivatives, let us

note that it is defined as

$$\frac{\delta F[u(x)]}{\delta u(y)} = \lim_{\epsilon \to 0} \frac{1}{\epsilon} \; (F[u(x)+\epsilon\delta(x-y)] - F[u(x)]) \quad (1.48)$$

Eq. (1.47) follows from this definition and the form of H in Eq. (1.46).]

With this choice of H, we note that $\frac{\delta H}{\delta u(x)}$ coincides with the quantity inside the bracket on the right hand side of Eq. (1.45). That is, we can write the KdV equation as

$$\frac{\partial u}{\partial t} = \frac{\partial}{\partial x} \frac{\delta H}{\delta u(x)} \quad (1.49)$$

On the other hand, for this equation to be Hamiltonian we must have

$$\frac{\partial u}{\partial t} = \{\dot{u}(x),H\} \quad (1.50)$$

Comparing with Eq. (1.49) we see that this requires

$$\{u(x),H\} = \frac{\partial}{\partial x} \frac{\delta H}{\delta u(x)}$$

or, $$\int_{-\infty}^{\infty} dy \; \frac{\delta H}{\delta u(y)} \; \{u(x),u(y)\} = \frac{\partial}{\partial x} \frac{\delta H}{\delta u(x)} \quad (1.51)$$

This determines the fundamental Poisson bracket to be

$$\{u(x),u(y)\} = \frac{\partial}{\partial x} \; \delta(x-y) \quad (1.52)$$

Being the derivative of a delta function, it is automatically antisymmetric. Before proving the Jacobi identity let us note parenthetically that this is the Poisson bracket structure associated with the Abelian current algebra in 1+1 dimension and hence must automatically satisfy the Jacobi identity. But let us work it out explicitly.

If we are given two functionals $F[u]$ and $G[u]$, then the choice of Eq. (1.52) as the fundamental Poisson bracket, would lead to a definition of the Poisson brackets as (see Eq. (1.20))

$$\{F[u],G[u]\}$$

$$= \int_{-\infty}^{\infty} dxdy \; \frac{\delta F}{\delta u(x)} \frac{\partial}{\partial x} \delta(x-y) \frac{\delta G}{\delta u(y)}$$

$$= \int_{-\infty}^{\infty} dx \; \frac{\delta F}{\delta u(x)} \frac{\partial}{\partial x} \frac{\delta G}{\delta u(x)} = - \int_{-\infty}^{\infty} dx \; \frac{\partial}{\partial x} \frac{\delta F}{\delta u(x)} \frac{\delta G}{\delta u(x)}$$

$$= \frac{1}{2} \int_{-\infty}^{\infty} dx \; \left(\frac{\delta F}{\delta u(x)} \frac{\partial}{\partial x} \frac{\delta G}{\delta u(x)} - \frac{\partial}{\partial x} \frac{\delta F}{\delta u(x)} \frac{\delta G}{\delta u(x)} \right) \qquad (1.53)$$

where we are assuming that $u(x,t)$ vanishes asymptotically fast enough so that the functional derivatives $\frac{\delta F}{\delta u(x)}$ and $\frac{\delta G}{\delta u(x)}$ also fall off rapidly at spatial infinity and integration by parts can be done without the need for any surface surface terms. Note now, from Eq. (1.53), that

$$\{H[u],\{F[u],G[u]\}\}$$

$$= - \int_{-\infty}^{\infty} dx \; \frac{\partial}{\partial x} \frac{\delta H}{\delta u(x)} \frac{\delta}{\delta u(x)} \{F[u],G[u]\} \qquad (1.54)$$

The functional derivative satisfies Leibniz's chain rule. Consequently, using the form of the Poisson bracket, Eq. (1.53), above and integrating by parts, Eq. (1.54) takes the form

$$\{H[u],\{F[u],G[u]\}\}$$

$$= - \int_{-\infty}^{\infty} dxdy \, \frac{\partial}{\partial x} \frac{\delta H}{\delta u(x)} \left(\frac{\delta^2 F}{\delta u(x) \delta u(y)} \frac{\partial}{\partial y} \frac{\delta G}{\delta u(y)} \right.$$

$$\left. - \frac{\partial}{\partial y} \frac{\delta F}{\delta u(y)} \frac{\delta^2 G}{\delta u(x) \delta u(y)} \right) \quad (1.55)$$

Given the expression, Eq. (1.55), it is straightforward to show that the Jacobi identity is satisfied. Namely,

$$\{H[u],\{F[u],G[u]\}\} + \{G[u],\{H[u],F[u]\}\} + \{F[u],\{G[u],H[u]\}\}$$

$$= - \int_{-\infty}^{\infty} dxdy \left[\frac{\partial}{\partial x} \frac{\delta H}{\delta u(x)} \frac{\delta^2 F}{\delta u(x) \delta u(y)} \frac{\partial}{\partial y} \frac{\delta G}{\delta u(y)} \right.$$

$$- \frac{\partial}{\partial x} \frac{\delta H}{\delta u(x)} \frac{\delta^2 G}{\delta u(x) \delta u(y)} \frac{\partial}{\partial y} \frac{\delta F}{\delta u(y)}$$

$$+ \frac{\partial}{\partial x} \frac{\delta G}{\delta u(x)} \frac{\delta^2 H}{\delta u(x) \delta u(y)} \frac{\partial}{\partial y} \frac{\delta F}{\delta u(y)}$$

$$- \frac{\partial}{\partial x} \frac{\delta G}{\delta u(x)} \frac{\delta^2 F}{\delta u(x) \delta u(y)} \frac{\partial}{\partial y} \frac{\delta H}{\delta u(y)}$$

$$+ \frac{\partial}{\partial x} \frac{\delta F}{\delta u(x)} \frac{\delta^2 G}{\delta u(x) \delta u(y)} \frac{\partial}{\partial y} \frac{\delta H}{\delta u(y)}$$

$$\left. - \frac{\partial}{\partial x} \frac{\delta F}{\delta u(x)} \frac{\delta^2 H}{\delta u(x) \delta u(y)} \frac{\partial}{\partial y} \frac{\delta G}{\delta u(y)} \right] = 0 \quad (1.56)$$

Thus this Poisson bracket structure satisfies Jacobi identity. In hindsight we could have shown it in a much simpler way. Namely, let us recall our discussion of Eqs. (1.12)-(1.17). As we have seen in Eq. (1.18), in the continuum case

$$f^{\mu\nu}(y) \rightarrow f(x,y,u(x)) = \frac{\partial}{\partial x} \delta(x-y) \qquad (1.57)$$

Therefore,

$$f_{\mu\nu}(y) \rightarrow f^{-1}(x,y,u(x)) = \epsilon(x-y) \qquad (1.58)$$

where $\epsilon(x-y)$ is the alternating step function defined to be

$$\epsilon(x-y) = (\theta(x-y) - \frac{1}{2}) \qquad (1.59)$$

The condition for Jacobi identity to hold, namely,

$$\partial_\mu f_{\nu\lambda} + \partial_\lambda f_{\mu\nu} + \partial_\nu f_{\lambda\mu} = 0$$

now becomes

$$\frac{\delta}{\delta u(z)} f^{-1}(x,y) + \frac{\delta}{\delta u(y)} f^{-1}(z,x) + \frac{\delta}{\delta u(x)} f^{-1}(y,z)$$

$$= \frac{\delta}{\delta u(z)} \epsilon(x-y) + \frac{\delta}{\delta u(y)} \epsilon(z-x) + \frac{\delta}{\delta u(x)} \epsilon(y-z) = 0 \qquad (1.60)$$

This is automatically true since the individual terms are independent of u and hence vanish identically.

Thus we see that the KdV equation is a Hamiltonian system with

$$H[u] = \int_{-\infty}^{\infty} dx \left(\frac{1}{3!} u^3(x) - \frac{1}{2} \left(\frac{\partial u}{\partial x}\right)^2 \right)$$

$$\{u(x),u(y)\} = \frac{\partial}{\partial x} \delta(x-y) \qquad (1.61)$$

so that

$$\frac{\partial u}{\partial t} = \{u(x),H\} = u \frac{\partial u}{\partial x} + \frac{\partial^3 u}{\partial x^3} \qquad (1.62)$$

In fact, let us note a peculiarity of the KdV equation. Namely, it is also a Hamiltonian system with a second choice of the Poisson bracket structure and a second Hamiltonian. For example, if we had chosen

$$\{u(x),u(y)\}_2 = \left(\frac{\partial^3}{\partial x^3} + \frac{1}{3} \left(\frac{\partial}{\partial x} u(x) + u(x) \frac{\partial}{\partial x} \right) \right) \delta(x-y) \quad (1.63)$$

and

$$H_2[u] = \int_{-\infty}^{\infty} dx \frac{1}{2} u^2(x) \qquad (1.64)$$

then

$$\frac{\partial u}{\partial t} = \{u(x), H_2\}_2$$

$$= \int_{-\infty}^{\infty} dy \, \frac{\delta H_2}{\delta u(y)} \, \{u(x), u(y)\}_2$$

$$= \int_{-\infty}^{\infty} dy \, u(y) \Big(\frac{\partial^3}{\partial x^3} + \frac{1}{3} \big(\frac{\partial}{\partial x} u(x) + u(x) \frac{\partial}{\partial x}\big)\Big) \delta(x-y)$$

$$= \Big(\frac{\partial^3}{\partial x^3} + \frac{1}{3} \big(\frac{\partial}{\partial x} u(x) + u(x) \frac{\partial}{\partial x}\big)\Big) u(x)$$

$$= \frac{\partial^3 u}{\partial x^3} + \frac{2}{3} u \frac{\partial u}{\partial x} + \frac{1}{3} u \frac{\partial u}{\partial x}$$

or, $\quad \dfrac{\partial u}{\partial t} = u \dfrac{\partial u}{\partial x} + \dfrac{\partial^3 u}{\partial x^3}$ $\hfill (1.65)$

Namely, this also yields the KdV equation. This Poisson bracket structure is also antisymmetric. Let us not go into the proof of the Jacobi identity but simply note here that this Poisson bracket structure is identical to the Virasoro algebra with a specific central charge. Hence the Jacobi identity must be satisfied.

Finally, let us conclude this chapter by showing that the KdV equation can be obtained as the Euler-Lagrange variation of a Lagrangian and point out some of the peculiarities. Consider the Lagrangian

$$L_{KdV} = \frac{1}{2} \int_{-\infty}^{\infty} dxdy \, u(x) \epsilon(x-y) \frac{\partial u(y)}{\partial t}$$

$$- \int_{-\infty}^{\infty} dx \Big(\frac{1}{3!} u^3(x) - \frac{1}{2} \big(\frac{\partial u}{\partial x}\big)^2\Big) \qquad (1.66)$$

The Euler-Lagrange equations following from this are

$$\int_{-\infty}^{\infty} dy\, \epsilon(x-y)\, \frac{\partial u(y)}{\partial t} = \frac{1}{2}\, u^2(x) + \frac{\partial^2 u}{\partial x^2}$$

or, $\quad \int_{-\infty}^{\infty} dy\, \frac{\partial}{\partial x}\, \epsilon(x-y)\, \frac{\partial u(y)}{\partial t} = \frac{\partial}{\partial x}\left(\frac{1}{2}\, u^2(x) + \frac{\partial^2 u}{\partial x^2}\right)$

or, $\quad \dfrac{\partial u(x)}{\partial t} = u\, \dfrac{\partial u}{\partial x} + \dfrac{\partial^3 u}{\partial x^3}$ (1.67)

This is, of course, the KdV equation. But the peculiarity of the Lagrangian is that it is nonlocal because of the presence of the term $\epsilon(x-y)$. In fact, one cannot write down a local Lagrangian in the variables $u(x)$ whose Euler-Lagrange equations would give the KdV equation.

References:

Abraham, R. and J. E. Marsden, Foundations of Mechanics, Benjamin/Cummings, 1978.

Arnold, V. I., Mathematical Methods of Classical Mechanics, Springer-Verlag, 1978.

Drazin, P. G., Solitons, Cambridge Univ. Press, 1983.

Eilenberger, G., Solitons, Springer-Verlag, 1983.

Korteweg, D. J. and G. de Vries, Phil. Mag. **39**, 422 (1895).

Kruskal, M. D., R. M. Miura, C. S. Gardner and N. Zabusky, J. Math. Phys. **11**, 952 (1970).

Lamb, G. L., Jr., Elements of Soliton Theory, John Wiley, 1980.

Magri, F., J. Math. Phys. <u>19</u>, 1156 (1978).

Newell, A. C., Solitons in Mathematical Physics, SIAM,
    1985.

Olshansky, M. A. and A. M. Perelomov, Phys. Rep. <u>71</u>, 315
    (1981).

Russell, J. S., Rep. 14th Meet. Brit. Assoc. Adv. Sci.,
    York 311 (1844).

Sudarshan, E. C. G. and N. Mukunda, Classical Dynamics:
    A Modern Approach, John Wiley, 1974.

Whitham, G. B., Linear and Nonlinear Waves, John Wiley,
    1974.

CHAPTER 2

PROPERTIES OF THE KDV SOLUTIONS

In the last chapter, we introduced the KdV equation as a Hamiltonian system. Before analyzing the integrability of this system, we would discuss various properties of the solutions of the KdV equation in this chapter. We would show that the KdV equation admits unique solutions with given initial conditions and that a class of solutions corresponds to solitons. We would also discuss scattering from a soliton potential to emphasize that it is reflectionless.

Uniqueness of Solutions:

Let us continue with the discussion of the KdV equation.

$$\frac{\partial u}{\partial t} = u \frac{\partial u}{\partial x} + \frac{\partial^3 u}{\partial x^3} \tag{2.1}$$

Let us recall that $u(x,t)$ is a function of time and one space coordinate. Given this equation, the first question we can ask is whether it admits a unique solution with given initial conditions. This can be easily determined as follows. Let us assume that $u(x,t)$ and $v(x,t)$ represent two solutions to the KdV equation satisfying the same initial conditions. That is,

$$\frac{\partial u}{\partial t} = u \frac{\partial u}{\partial x} + \frac{\partial^3 u}{\partial x^3} \qquad (2.2)$$

and

$$\frac{\partial v}{\partial t} = v \frac{\partial v}{\partial x} + \frac{\partial^3 v}{\partial x^3} \qquad (2.3)$$

with

$$u(x,0) = v(x,0) = f(x) \qquad (2.4)$$

If this is true, then subtracting Eq. (2.3) from Eq. (2.2), we obtain

$$\frac{\partial}{\partial t} (u-v) = u \frac{\partial u}{\partial x} - v \frac{\partial v}{\partial x} + \frac{\partial^3 (u-v)}{\partial x^3}$$

$$= u \frac{\partial (u-v)}{\partial x} + (u-v) \frac{\partial v}{\partial x} + \frac{\partial^3 (u-v)}{\partial x^3}$$

$$\text{or,} \quad \frac{\partial w}{\partial t} = u \frac{\partial w}{\partial x} + w \frac{\partial v}{\partial x} + \frac{\partial^3 w}{\partial x^3} \qquad (2.5)$$

Here we have defined

$$w = u(x,t) - v(x,t) \qquad (2.6)$$

Furthermore, let us remind ourselves that we are all along assuming the functions u and v to be vanishing at spatial infinity. Now, multiplying the above equation with w and integrating over x, we obtain

$$\frac{d}{dt} \int_{-\infty}^{\infty} dx \, \frac{1}{2} \, w^2 = \int_{-\infty}^{\infty} dx \, w^2 (\frac{\partial v}{\partial x} - \frac{1}{2} \frac{\partial u}{\partial x}) \qquad (2.7)$$

Here we have used the fact that u and v and, therefore, w fall off to zero at spatial infinity. Let us next define

$$E(t) = \int_{-\infty}^{\infty} dx \, \frac{1}{2} \, w^2(x,t) \qquad (2.8)$$

and

$$m = 2\max \left| \frac{\partial v}{\partial x} - \frac{1}{2} \frac{\partial u}{\partial x} \right| \qquad (2.9)$$

We can, then, write Eq. (2.7) as the inequality

$$\frac{dE(t)}{dt} \leq mE(t)$$

$$\text{or,} \quad E(t) \leq E(0)e^{mt} \qquad (2.10)$$

Clearly, since E(t) by definition is positive semidefinite, it follows from Eq. (2.10) that E(t) must vanish if E(0) vanishes. That is,

$$E(0) = 0$$

would simply   (2.11)

$$E(t) = 0$$

Now since we know from Eq. (2.4) that u and v satisfy the same initial data, namely,

$$u(x,0) = v(x,0) = f(x)$$

it follows that

$$w(x,0) = u(x,0) - v(x,0) = 0 \qquad\qquad (2.12)$$

This implies that

$$E(0) = \frac{1}{2} \int_{-\infty}^{\infty} dx\ w^2(x,0) = 0 \qquad\qquad (2.13)$$

Consequently, from Eq. (2.11) it follows that

$$E(t) = 0 \qquad\qquad (2.14)$$

On the other hand, since

$$E(t) = \frac{1}{2} \int_{-\infty}^{\infty} dx\ w^2(x,t)$$

Eq. (2.14) implies that

$$w(x,t) = 0$$

$$\text{or,}\quad u(x,t) = v(x,t) \qquad\qquad (2.15)$$

This proves that if there are two solutions of the KdV
equation satisfying the same initial data, they must be the
same. Therefore, the KdV equation admits unique solutions
with given initial conditions.

Travelling Waves:

   Next, let us examine the travelling wave solutions of
the KdV equation. Namely, let us assume that

$$u(x,t) = u(x+ct) = f(x+ct) \qquad \text{with} \quad c > 0 \quad (2.16)$$

where c is the speed of the travelling wave which satisfies
the KdV equation. Then putting this back into the KdV
equation, Eq. (2.1), we obtain

$$\frac{\partial u}{\partial t} = u \frac{\partial u}{\partial x} + \frac{\partial^3 u}{\partial x^3}$$

$$\text{or,} \quad c \frac{\partial u}{\partial x} = u \frac{\partial u}{\partial x} + \frac{\partial^3 u}{\partial x^3} \qquad\qquad (2.17)$$

Since Eq. (2.17) involves only the x-derivatives, we can set
t=0. Then the equation becomes

$$c \frac{du}{dx} = \frac{d}{dx} \left(\frac{1}{2} u^2 \right) + \frac{d}{dx} \left(\frac{d^2 u}{dx^2} \right) \qquad\qquad (2.18)$$

This implies that

$$\frac{d^2u}{dx^2} + \frac{1}{2} u^2 - cu = \text{constant}. \tag{2.19}$$

From the boundary condition that $u(x,t)$ must vanish at spatial infinity, it follows that the constant on the right hand side of Eq. (2.19) must vanish. Thus

$$\frac{d^2u}{dx^2} + \frac{1}{2} u^2 - cu = 0 \tag{2.20}$$

Multiplying Eq. (2.20) with $\frac{du}{dx}$, we obtain

$$\frac{du}{dx} \frac{d^2u}{dx^2} + \frac{1}{2} \frac{du}{dx} u^2 - c \frac{du}{dx} u = 0$$

$$\text{or,} \quad \frac{1}{2} \frac{d}{dx} (\frac{du}{dx})^2 + \frac{1}{6} \frac{du^3}{dx} - \frac{c}{2} \frac{du^2}{dx} = 0$$

$$\text{or,} \quad \frac{1}{2} (\frac{du}{dx})^2 + \frac{1}{6} u^3 - \frac{c}{2} u^2 = 0 \tag{2.21}$$

Note that we have again set the constant to zero to conform to the boundary conditions. If we now choose the condition

$$\frac{du}{dx} = 0 \quad \text{for} \quad x = 0 \tag{2.22}$$

the solution can be obtained in the closed form to be

$$u(x) = 3c \ \text{sech}^2 \ \frac{\sqrt{c}}{2} \ x \tag{2.23}$$

Furthermore, restoring the time variable, we obtain

$$u(x,t) = 3c \ \text{sech}^2 \ \frac{\sqrt{c}}{2} \ (x+ct) \tag{2.24}$$

This shows that the KdV equation possesses travelling wave solutions.

Solitons:

We now readily see that the travelling wave of Eq. (2.24) possesses the following interesting properties.

i) $u(x,t) \rightarrow 0$ as $x \rightarrow \pm \infty$ and furthermore it is fairly localized.

ii)   The wave travels only to the left. (In fact, if we change $c \rightarrow -c$, the wave which would be moving to the right becomes oscillatory.)

iii)   The amplitude of the wave is directly proportional to its velocity. That is, the taller the wave the faster it moves. This actually is observed in the case of solitary water waves.

iv)   Finally, the most important of all is that the wave has no dispersion. That is, it maintains its shape as it moves. To see this, note that since

$$u(x,t) = f(x+ct)$$

each Fourier mode would have the form

$$u_k(x,t) = e^{ik(x+ct)}\tilde{f}(k)$$

and, therefore, we can readily identify the energy associated with each wave component to be

$$E_k = \omega_k = ck \qquad\qquad (2.25)$$

where k is the wave number associated with the Fourier mode. Consequently,

$$\frac{E_k}{k} = \frac{\omega_k}{k} = c = constant = \frac{dE_k}{dk} \qquad\qquad (2.26)$$

Thus we see that each component of the wave moves with the same constant phase velocity which is also equal to the group velocity. Consequently, the motion is nondispersive.

This is, of course, the definition of a soliton solution. Namely, it is a wave without any dispersion and consequently, one which maintains its shape as it moves. This analysis, therefore, shows that the KdV equation admits soliton solutions. (This should come as no surprise since this equation was formulated to describe solitary water waves.) To emphasize the significance of the nondispersive nature of this solution, let us in fact compare this with the free particle solution of the Schrödinger equation. Namely, we know that the free Schrödinger equation

$$i\hbar \frac{\partial\psi(x,t)}{\partial t} = - \frac{\hbar^2}{2m} \frac{\partial^2\psi(x,t)}{\partial x^2} \qquad (2.27)$$

has the closed form solution given by

$$\psi(x,t) = \frac{N}{\left(a^2 + \frac{i\hbar t}{m}\right)^{1/2}} e^{-x^2/2\left(a^2 + \frac{i\hbar t}{m}\right)} \qquad (2.28)$$

Here N is the normalization constant and "a" is a constant which depends on the initial condition. We can calculate from Eq. (2.28) the mean width associated with the free particle and it has the form

$$(\Delta x)^2 = \frac{1}{2} \left(a^2 + \frac{\hbar^2 t^2}{m^2 a^2}\right) \qquad (2.29)$$

Clearly, the mean width increases with time and consequently the wave packet becomes spread out with time. That is, even the free particle solution of the linear Schrödinger equation is dispersive in nature and the reason behind this can be traced to the following fact. We know that in this case

$$E_k = \omega_k = \frac{\hbar^2 k^2}{2m} \qquad (2.30)$$

Consequently,

$$v_{ph} = \frac{E_k}{k} = \frac{\omega_k}{k} = \frac{\hbar^2 k}{2m} = \text{phase velocity} \qquad (2.31)$$

$$v_g = \frac{dE_k}{dk} = \frac{d\omega_k}{dk} = \frac{\hbar^2 k}{m} = \text{group velocity} \qquad (2.32)$$

That is, each Fourier component of the wave packet moves with a different phase velocity proportional to its wave number and that the phase velocity is different from the group velocity. This leads to dispersion as shorter waves move faster than the longer waves and gives rise to the change in the shape of the wave packet.

Let us drop the nonlinear term in the KdV equation and analyze the dispersion relation for

$$\frac{\partial u}{\partial t} = \frac{\partial^3 u}{\partial x^3} \qquad (2.33)$$

Clearly this linear equation would lead to a dispersion relation of the form

$$E_k = \omega_k = k^3 \qquad (2.34)$$

so that for any solution of this linear equation we would obtain

$$v_{ph} = \frac{E_k}{k} = \frac{\omega_k}{k} = k^2 \qquad (2.35)$$

$$v_g = \frac{dE_k}{dk} = \frac{d\omega_k}{dk} = 3k^2 \qquad (2.36)$$

That is, if we neglect the nonlinear term in the KdV equation, the solutions are also dispersive as in the case of the free Schrödinger equation. This implies, therefore, that it is the nonlinear term in the KdV equation which is responsible for the nondispersive nature or the soliton nature of the solutions.

Quantum Mechanics of the Soliton Solutions:

Since soliton solutions are universal in integrable models, let us digress a little and study the quantum mechanics of such potentials.

Let us first discuss some general features of one dimensional Hamiltonians. Let the Hamiltonian

$$H = -\frac{1}{2} D^2 + \frac{1}{2} U(x) \qquad (2.37)$$

have the eigenfunctions $\phi(x)$ satisfying

$$H\phi(x) = \lambda\phi(x)$$

$$\text{or,} \quad \left(-\frac{1}{2} D^2 + \frac{1}{2} U(x)\right) \phi(x) = \lambda\phi(x) \qquad (2.38)$$

Here we have set the mass to be unity and have denoted $\frac{\partial}{\partial x}$ by D. Note that if $\phi(x)$ is non-trivial then we can define an operator

$$A(x) = \frac{1}{\sqrt{2}} \phi(x)D\phi^{-1}(x)$$

$$= \frac{1}{\sqrt{2}} \, \phi(x) \, \left(- \, \phi^{-2}(D\phi(x)) + \phi^{-1}(x)D \right)$$

or, $A(x) = \frac{1}{\sqrt{2}} \left(D - \phi^{-1}(x)(D\phi(x)) \right)$ (2.39)

Similarly, we can also define the operator

$$A^{+}(x) = - \, \frac{1}{\sqrt{2}} \, \phi^{-1}(x)D\phi(x)$$

$$= - \, \frac{1}{\sqrt{2}} \, \phi^{-1}(x) \left((D\phi(x)) + \phi(x)D \right)$$

$$= - \, \frac{1}{\sqrt{2}} \left(D + \phi^{-1}(x)(D\phi(x)) \right)$$ (2.40)

If we assume that $\phi(x)$ is real, we see that $A^{+}(x)$ is indeed the formal adjoint of $A(x)$. Now

$A^{+}(x)A(x)$

$$= - \, \frac{1}{\sqrt{2}} \left(D + \phi^{-1}(x)(D\phi(x)) \right) \frac{1}{\sqrt{2}} \left(D - \phi^{-1}(x)(D\phi(x)) \right)$$

$$= - \, \frac{1}{2} \left(D^{2} + \phi^{-1}(x)(D\phi(x))D - D\phi^{-1}(x)(D\phi(x)) \right.$$

$$\left. - \, \phi^{-1}(x)(D\phi(x))\phi^{-1}(x)(D\phi(x)) \right)$$

$$= - \, \frac{1}{2} \left(D^{2} + \phi^{-1}(x)(D\phi(x))D + \phi^{-2}(D\phi(x))^{2} - \phi^{-1}(x)(D^{2}\phi(x)) \right.$$

$$\left. - \, \phi^{-1}(x)(D\phi(x))D - \phi^{-2}(x)(D\phi(x))^{2} \right)$$

$$= - \frac{1}{2} \left(D^2 - \phi^{-1}(x)(D^2\phi(x)) \right)$$

or, $A^+(x)A(x) = \left(- \frac{1}{2} D^2 + \phi^{-1}(x)(\frac{1}{2} D^2\phi(x)) \right)$ (2.41)

But $\phi(x)$ is an eigenstate of the Hamiltonian and, therefore, if we use Eq. (2.38), we obtain

$$A^+(x)A(x) = \left(- \frac{1}{2} D^2 + \phi^{-1}(x)(\frac{1}{2} U(x)\phi(x) - \lambda\phi(x)) \right)$$

$$= \left(- \frac{1}{2} D^2 + \frac{1}{2} U(x) - \lambda \right)$$

or, $A^+(x)A(x) = (H - \lambda I)$ (2.42)

This shows that the one dimensional operator $H-\lambda I$ can always be factorized into the above form with

$$A(x) = \frac{1}{\sqrt{2}} (D - W(x))$$

and

$$A^+(x) = - \frac{1}{\sqrt{2}} (D + W(x))$$ (2.43)

where $W(x) = \phi^{-1}(x)(D\phi(x))$ and $\phi(x)$ is the eigenstate of the Hamiltonian with eigenvalue λ. Note that this identification induces a Riccati type transformation between $U(x)$ and $W(x)$, namely,

$$U(x) - 2\lambda = W^2(x) + (DW(x))$$ (2.44)

Thus we see that given any eigenfunction $\phi(x)$ of H, the operator H - λI factorizes. Conversely, given any constant λ and a given potential, if we can find a generalized Riccati type relation of the kind in Eq. (2.44), then the operator H - λI can be factorized. Let us now study the properties of the operators A(x) and $A^+(x)$. Note from Eq. (2.39) that

$$A(x) = \frac{1}{\sqrt{2}} \phi(x) D \phi^{-1}(x)$$

Consequently,

$$A(x)\phi(x) = \frac{1}{\sqrt{2}} \phi(x) D(\phi^{-1}(x)\phi(x)) = 0$$ (2.45)

so that $\phi(x)$ is annihilated by A(x). In terms of W(x), this equation reads as

$$\frac{1}{\sqrt{2}} (D - W(x))\phi(x) = 0$$

$$\text{or,} \quad \frac{d\phi(x)}{dx} = W(x)\phi(x)$$

$$\text{or,} \quad \phi(x) = e^{\int^{x} W(x')dx'}$$ (2.46)

Hence if we know the form of W(x), whether a function $\phi(x)$ exists which would lead to factorization depends on whether the solution above is normalizeable or not. We note also from Eq. (2.40) that

$$A^+(x)\phi^{-1}(x) = -\frac{1}{\sqrt{2}}\ \phi^{-1}(x)D(\phi(x)\phi^{-1}(x)) = 0 \quad (2.47)$$

Clearly, therefore, $\phi(x)$ and $\phi^{-1}(x)$ are eigenstates of zero eigenvalue for the operators $A(x)$ and $A^+(x)$ respectively.

Let us note next that we can form a second Hermitian bilinear from $A(x)$ and $A^+(x)$. Namely,

$$A(x)A^+(x)$$

$$= \frac{1}{\sqrt{2}}\left(D - \phi^{-1}(x)(D\phi(x))\right)(-\frac{1}{\sqrt{2}})\left(D + \phi^{-1}(x)(D\phi(x))\right)$$

$$= -\frac{1}{2}\left(D^2 - \phi^{-1}(x)(D\phi(x))D + D\phi^{-1}(x)(D\phi(x))\right.$$

$$\left. - \phi^{-1}(x)(D\phi(x))\phi^{-1}(x)(D\phi(x))\right)$$

$$= -\frac{1}{2}\left(D^2 - \phi^{-1}(x)(D\phi(x))D - \phi^{-2}(x)(D\phi(x))^2\right.$$

$$\left. + \phi^{-1}(x)(D^2\phi(x)) + \phi^{-1}(x)(D\phi(x))D - \phi^{-2}(x)(D\phi(x))^2\right)$$

$$= -\frac{1}{2}\left(D^2 + \phi^{-1}(x)(D^2\phi(x)) - 2\phi^{-2}(x)(D\phi(x))^2\right) \quad (2.48)$$

If we now use the identity

$$D^2\ell n\phi(x) = D\left(\phi^{-1}(x)(D\phi(x))\right)$$

$$= -\phi^{-2}(x)(D\phi(x))^2 + \phi^{-1}(x)(D^2\phi(x)) \quad (2.49)$$

then we can write Eq. (2.48) as

$$A(x)A^+(x)$$

$$= -\frac{1}{2}\left(D^2 + \phi^{-1}(x)(D^2\phi(x)) + 2(D^2\ln\phi(x))\right.$$

$$\left. - 2\phi^{-1}(x)(D^2\phi(x))\right)$$

$$= -\frac{1}{2}(D^2 - \phi^{-1}(x)(D^2\phi(x)) + 2(D^2\ln\phi(x)))$$

$$= \left(-\frac{1}{2}D^2 + \phi^{-1}(x)(\frac{1}{2}D^2\phi(x)) - (D^2\ln\phi(x))\right)$$

$$= \left(-\frac{1}{2}D^2 + \phi^{-1}(x)(\frac{1}{2}U(x) - \lambda)\phi(x) - (D^2\ln\phi(x))\right)$$

$$= \left(-\frac{1}{2}D^2 + \frac{1}{2}U(x) - \lambda - (D^2\ln\phi(x))\right)$$

or, $A(x)A^+(x) = -\frac{1}{2}D^2 + \frac{1}{2}\tilde{U}(x) - \lambda = \tilde{H} - \lambda I$ (2.50)

where

$$\tilde{U}(x) = U(x) - 2(D^2\ln\phi(x))$$

That is, if

$$A^+(x)A(x) = H_+ = H - \lambda I = -\frac{1}{2}D^2 + \frac{1}{2}U(x) - \lambda \qquad (2.51)$$

then

$$A(x)A^+(x) = H_- = \tilde{H} - \lambda I = -\frac{1}{2} D^2 + \frac{1}{2} \tilde{U}(x) - \lambda \qquad (2.52)$$

We remark here parentetically that H_+ and H_- can be thought of as supersymmetric partners of each other.

The significance of going through this formalism is to recognize that the Hamiltonians H_+ and H_- are almost isospectral. That is, they have almost the same eigenvalues. The easiest way to see this is to note that if

$$H_+\psi = A^+(x)A(x)\psi = \epsilon\psi \qquad (2.53)$$

then

$$A(x)A^+(x)A(x)\psi = \epsilon A(x)\psi$$

$$\text{or,} \quad H_-(A(x)\psi) = \epsilon(A(x)\psi) \qquad (2.54)$$

That is, if $\psi(x)$ is an eigenstate of H_+ with a nontrivial eigenvalue ϵ, then $A(x)\psi(x)$ is an eigenstate of H_- with the same eigenvalue. That is, the eigenstates of H_+ and H_- are paired except for the state which satisfies

$$A(x)\psi(x) = 0 \qquad (2.55)$$

This is why we say that the two Hamiltonians are almost isospectral. Note that in terms of the $W(x)$'s we can write

$$H_+ = A^+(x)A(x)$$

$$= \left(-\frac{1}{2} D^2 + \frac{1}{2} W^2(x) + \frac{1}{2} (DW(x)) \right)$$

(2.56)

$$H_- = A(x)A^+(x)$$

$$= \left(-\frac{1}{2} D^2 + \frac{1}{2} W^2(x) - \frac{1}{2} (DW(x)) \right)$$

Let us next apply this analysis to the potential

$$\frac{1}{2} U_n(x) = -\frac{n(n+1)}{2} \text{sech}^2 x$$

(2.57)

to determine the spectrum of the Hamiltonian

$$H_n = -\frac{1}{2} D^2 - \frac{n(n+1)}{2} \text{sech}^2 x$$

(2.58)

Here we are assuming n to be a positive integer. Let us note that if we choose

$$W_n(x) = -n \tanh x$$

(2.59)

then

$$W_n^2(x) + DW_n(x)$$

$$= n^2 \tanh^2 x - n\,\text{sech}^2 x$$

$$= n^2(1 - \text{sech}^2 x) - n\,\text{sech}^2 x$$

$$= n^2 - n(n+1)\text{sech}^2 x$$

so that, from Eqs. (2.56) and (2.58), we obtain

$$H_{n+} = -\frac{1}{2} D^2 + \frac{1}{2} W_n^2(x) + \frac{1}{2} (DW_n(x))$$

$$= -\frac{1}{2} D^2 - \frac{n(n+1)}{2} \text{sech}^2 x + \frac{n^2}{2}$$

$$= H_n + \frac{n^2}{2} \tag{2.60}$$

Similarly, we have

$$H_{n-} = -\frac{1}{2} D^2 + \frac{1}{2} W_n^2(x) - \frac{1}{2} (DW_n(x))$$

$$= -\frac{1}{2} D^2 + \frac{n^2}{2} \tanh^2 x + \frac{n}{2} \text{sech}^2 x$$

$$= -\frac{1}{2} D^2 + \frac{n^2}{2} (1-\text{sech}^2 x) + \frac{n}{2} \text{sech}^2 x$$

$$= -\frac{1}{2} D^2 - \frac{n(n-1)}{2} \text{sech}^2 x + \frac{n^2}{2}$$

$$= H_{n-1} + \frac{n^2}{2} \tag{2.61}$$

Now since H_{n+} and H_{n-} are almost isospectral, it follows from Eqs. (2.60) and (2.61) that H_n and H_{n-1} are also almost isospectral. The only eigenvalue that is not shared by H_n and H_{n-1} is given by

$$\left(H_n + \frac{n^2}{2} \right) \phi_n(x) = 0$$

$$\text{or,} \quad E_n = -\frac{n^2}{2} \tag{2.62}$$

Similarly, we can show that H_{n-1} and H_{n-2} have all their eigenvalues paired except the one for which

$$\left(H_{n-1} + \frac{(n-1)^2}{2} \right) \phi_{n-1} = 0$$

$$\text{or,} \quad E_{n-1} = -\frac{(n-1)^2}{2} \tag{2.63}$$

One could follow this down to lower values of n until n=0 which corresponds to the free Hamiltonian. Thus we see that the discrete energy eigenvalues of H_n are given by

$$E_k = -\frac{k^2}{2} \qquad \text{with} \quad k = 1, 2, \ldots . n \tag{2.64}$$

Consequently, we see that the potential

$$\frac{1}{2} U_n(x) = -\frac{n(n+1)}{2} \operatorname{sech}^2 x$$

has exactly n bound states. It is worth comparing this with the form of the solution (Eq. 2.23) we had found earlier for the KdV equation. As we will show in Chapter 4, Eq. (2.23) with c=4 supports only one bound state and describes a single soliton of the KdV equation.

Let us also note that the eigenstates of H_n and H_{n-1} are related through the operator $A_n(x)$ and $A_n^+(x)$ which has the form

$$A_n(x) = \frac{1}{\sqrt{2}} (D - W_n(x)) = \frac{1}{\sqrt{2}} \left(\frac{\partial}{\partial x} - W_n(x) \right)$$

$$\text{(2.65)}$$

$$A_n^+(x) = -\frac{1}{\sqrt{2}} (D + W_n(x)) = -\frac{1}{\sqrt{2}} \left(\frac{\partial}{\partial x} + W_n(x) \right)$$

Furthermore, since all the H_n's are isospectral with H_0 whose eigenfunctions are plane waves, all the bound states of H_n can be obtained from them by successive application of appropriate $A_n^+(x)$'s. Let us note that H_0 is the free Hamiltonian and hence does not give rise to any reflection. Furthermore, the operators

$$A_n^+(x) = -\frac{1}{\sqrt{2}} \left(\frac{\partial}{\partial x} + W_n(x) \right)$$

do not mix up the plane waves $e^{\pm ikx}$. Consequently, the Hamiltonians H_n for any positive integer n do not lead to any reflection. Consequently, these potentials are termed reflectionless. We should have, of course, anticipated this from the fact that these potentials correspond to solitons and solitons just pass through each other. As we will see later, this property is invaluable in finding explicit solutions.

References:

Adler, M. and J. Moser, Comm. Math. Phys. <u>61</u>, 1 (1978).

Crum, M. M., Quart. J. Math. Ser2 <u>6</u>, 121 (1955).

Deift, P., Duke Math. J. <u>45</u>, 267 (1978).

Gardner, C. S., J. M. Greene, M. D. Kruskal and R. M.
        Miura, Comm. Pure Appl. Math. <u>27</u>, 97 (1974).

Gendenshtein, L. E., JETP Lett. <u>38</u>, 356 (1983).

Kruskal, M. in Dynamical Systems, Theory and Applica-
        tions, Ed. J. Moser, Springer-Verlag, 1974.

Lax, P. D., Comm. Pure Appl. Math. <u>21</u>, 467 (1968).

Lighthill, J., Waves in Fluids, Cambridge Univ. Pres,
        1978.

CHAPTER 3

INTEGRABILITY OF THE KDV EQUATION

So far we have seen that the KdV equation admits unique solutions with any given initial condition and that solitons are a class of solutions that satisfy the KdV equation. On the other hand, we have also argued that soliton solutions imply the existence of an infinite number of conserved quantities. In this chapter we will show that the KdV equation does possess an infinite number of conserved quantities which are in involution and, therefore, the KdV equation is integrable.

Conserved Quantities:

Let us begin with a brief and general discussion about conserved quantities. We recall that if $Q[u]$ is conserved, then

$$\frac{dQ[u]}{dt} = \{Q[u], H\} = 0 \qquad (3.1)$$

Furthermore, let us explicitly represent

$$Q[u] = \int_{-\infty}^{\infty} dx\ \rho[u(x,t)] \qquad (3.2)$$

Then

$$\frac{dQ[u]}{dt} = \int_{-\infty}^{\infty} dx \; \frac{\partial \rho[u(x,t)]}{\partial t} = 0 \qquad (3.3)$$

would imply the existence of a continuity equation of the form

$$\frac{\partial \rho[u(x,t)]}{\partial t} + \frac{\partial j[u(x,t)]}{\partial x} = 0 \qquad (3.4)$$

Thus, whereas the integrated charges are independent of time (constants of motion), the densities themselves would satisfy continuity equations.

Given this we immediately recognize that the evolution equation, Eq. (1.37), namely, the KdV equation itself, is already in the form of a continuity equation. That is, since

$$\frac{\partial u}{\partial t} = \frac{\partial}{\partial x} \left(\frac{1}{2} u^2 + \frac{\partial^2 u}{\partial x^2} \right) \qquad (3.5)$$

comparing with Eq. (3.4), we can immediately identify

$$\rho_0[u(x,t)] = u(x,t)$$

$$\qquad (3.6)$$

$$j_0 = -\left(\frac{1}{2} u^2 + \frac{\partial^2 u}{\partial x^2} \right)$$

Thus we recognize that

$$Q_0 = H_0 = \int_{-\infty}^{\infty} dx \rho_0 [u(x,t)] = \int_{-\infty}^{\infty} dx \ u(x,t)$$

is a constant of motion. For later use, however, let us redefine

$$\rho_0 [u(x,t)] = 3u(x,t)$$

(3.7)

$$H_0 = \int_{-\infty}^{\infty} dx \ \rho_0 [u(x,t)] = 3 \int_{-\infty}^{\infty} dx \ u(x,t)$$

Next, let us note that since

$$\frac{\partial u}{\partial t} = u \frac{\partial u}{\partial x} + \frac{\partial^3 u}{\partial x^3}$$

therefore,

$$u \frac{\partial u}{\partial t} = u^2 \frac{\partial u}{\partial x} + u \frac{\partial^3 u}{\partial x^3}$$

or, $\frac{\partial}{\partial t} (\frac{1}{2} u^2) = \frac{\partial}{\partial x} (\frac{1}{3} u^3) + \frac{\partial}{\partial x} \left(u \frac{\partial^2 u}{\partial x^2} - \frac{1}{2} (\frac{\partial u}{\partial x})^2 \right)$

or, $\frac{\partial}{\partial t} (\frac{1}{2} u^2) = \frac{\partial}{\partial x} \left(\frac{1}{3} u^3 - \frac{1}{2} (\frac{\partial u}{\partial x})^2 + u \frac{\partial^2 u}{\partial x^2} \right)$ (3.8)

Consequently, we have a second continuity equation if we identify

$$\rho_1[u(x,t)] = \frac{1}{2} u^2$$

$$j_1 = -\frac{1}{3} u^3 + \frac{1}{2} \left(\frac{\partial u}{\partial x}\right)^2 - u \frac{\partial^2 u}{\partial x^2} \qquad (3.9)$$

A second constant of motion, therefore, is given by

$$H_1 = \int_{-\infty}^{\infty} dx \frac{1}{2} (u(x,t))^2 \qquad (3.10)$$

Furthermore, we note from Eq. (1.61) that the Hamiltonian for the KdV equation has the form

$$H_{KdV} = \int_{-\infty}^{\infty} dx \left(\frac{1}{3!} u^3 - \frac{1}{2} \left(\frac{\partial u}{\partial x}\right)^2\right) \qquad (3.11)$$

This must also be a constant of motion since

$$\frac{dH_{KdV}}{dt} = \{H_{KdV}, H_{KdV}\} = 0$$

Hence we can identify

$$H_2 = H_{KdV} = \int_{-\infty}^{\infty} dx \left(\frac{1}{3!} u^3 - \frac{1}{2} \left(\frac{\partial u}{\partial x}\right)^2\right) \qquad (3.12)$$

Let us note that with the fundamental Poisson bracket given by

$$\{u(x,t),u(y,t)\} = \frac{\partial}{\partial x}\, \delta(x-y) \qquad\qquad (3.13)$$

we can identify H_2 with the generator of time translations and, therefore, the energy since

$$\{u(x,t),H_2\} = \frac{\partial u}{\partial t} \qquad\qquad (3.14)$$

Similarly, H_1 can be thought of as the momentum with the Poisson bracket of Eq. (1.61) since it generates space translations. Namely,

$$\{u(x,t),H_1\} = \int_{-\infty}^{\infty} dy\, \{u(x,t),\, \tfrac{1}{2}\, u^2(y,t)\}$$

$$= \int_{-\infty}^{\infty} dy\, u(y,t)\{u(x,t),u(y,t)\}$$

$$= \int_{-\infty}^{\infty} dy\, u(y,t)\, \frac{\partial}{\partial x}\, \delta(x-y)$$

$$\text{or,} \quad \{u(x,t),H_1\} = \frac{\partial u(x,t)}{\partial x} \qquad\qquad (3.15)$$

These are the only conserved quantities that we can think of from naive symmetry considerations. Historically, therefore, when a fourth conserved quantity was discovered and was found not to be related to any naive symmetry, it caused a lot of interest. Moreover, there were some conjectures that the number of conserved quantities can only be a maximum of seven. Therefore, when Miura had found eleven conserved quantities all bets were off and in fact

Kruskal predicted that there must be an infinite number of them.

Let me emphasize that all these conserved quantities, namely eleven of them, were constructed by brute force. Various systematic approaches to the problem have developed since and we will discuss the method which is simple and elegant.

The Miura Transformation:

To construct, systematically, the conserved quantities of the KdV equation, let us introduce a second equation related to the KdV equation known as the modified KdV equation. This is given by

$$\frac{\partial v}{\partial t} = v^2 \frac{\partial v}{\partial x} + \frac{\partial^3 v}{\partial x^3} \tag{3.16}$$

It is called the modified KdV equation (MKdV equation) because it can be obtained from the KdV equation by the Riccati transformation

$$u(x,t) = v^2(x,t) + i\sqrt{6} \frac{\partial v}{\partial x} \tag{3.17}$$

This can be readily checked. Since

$$\frac{\partial u}{\partial t} = u \frac{\partial u}{\partial x} + \frac{\partial^3 u}{\partial x^3}$$

this implies

$$2v \frac{\partial v}{\partial t} + i\sqrt{6} \frac{\partial^2 v}{\partial x \partial t} = \left(v^2 + i\sqrt{6} \frac{\partial v}{\partial x} \right) \left(2v \frac{\partial v}{\partial x} + i\sqrt{6} \frac{\partial^2 v}{\partial x^2} \right)$$

$$+ 2v \frac{\partial^3 v}{\partial x^3} + 4 \frac{\partial v}{\partial x} \frac{\partial^2 v}{\partial x^2} + 2 \frac{\partial v}{\partial x} \frac{\partial^2 v}{\partial x^2} + i\sqrt{6} \frac{\partial^4 v}{\partial x^4}$$

or, $$\left(2v + i\sqrt{6} \frac{\partial}{\partial x} \right) \frac{\partial v}{\partial t} = 2v^3 \frac{\partial v}{\partial x} + i\sqrt{6} \, v^2 \frac{\partial^2 v}{\partial x^2}$$

$$+ 2i\sqrt{6} \, v \left(\frac{\partial v}{\partial x} \right)^2 - 6 \frac{\partial v}{\partial x} \frac{\partial^2 v}{\partial x^2} + 2v \frac{\partial^3 v}{\partial x^3}$$

$$+ 6 \frac{\partial v}{\partial x} \frac{\partial^2 v}{\partial x^2} + i\sqrt{6} \frac{\partial^4 v}{\partial x^4}$$

or, $$\left(2v + i\sqrt{6} \frac{\partial}{\partial x} \right) \frac{\partial v}{\partial t} = 2v^3 \frac{\partial v}{\partial x} + i\sqrt{6} \, v^2 \frac{\partial^2 v}{\partial x^2}$$

$$+ 2i\sqrt{6} \, v \left(\frac{\partial v}{\partial x} \right)^2 + 2v \frac{\partial^3 v}{\partial x^3} + i\sqrt{6} \frac{\partial^4 v}{\partial x^4}$$

$$\left(2v + i\sqrt{6} \frac{\partial}{\partial x} \right) \frac{\partial v}{\partial t} = \left(2v + i\sqrt{6} \frac{\partial}{\partial x} \right) \left(v^2 \frac{\partial v}{\partial x} + \frac{\partial^3 v}{\partial x^3} \right) \quad (3.18)$$

This shows that any solution of the MKdV equation automatically gives a solution of the KdV equation through the Riccati relation of Eq. (3.17). The MKdV equation is also known to be integrable and hence the conserved quantities of the MKdV equation also give rise to the conserved quantities of the KdV equation.

 In spite of the fact that the MKdV equation is derived
from the KdV equation, the symmetries of the two systems are
not the same. In fact whereas the KdV equation is Galilean
invariant (see Eq. (1.42)), the MKdV equation is not. One
can readily check that under

$$t \rightarrow t$$

$$x \rightarrow x + \frac{3}{2\epsilon^2} t$$

$$u \rightarrow u + \frac{3}{2\epsilon^2} \qquad\qquad (3.19)$$

$$v \rightarrow \frac{\epsilon}{\sqrt{6}} v + \frac{\sqrt{6}}{2\epsilon}$$

the KdV equation is invariant but the MKdV equation is not.
In fact, under this transformation

$$\frac{\partial v}{\partial t} = v^2 \frac{\partial v}{\partial x} + \frac{\partial^3 v}{\partial x^3} \quad \longrightarrow$$

$$\frac{\epsilon}{\sqrt{6}} \left(\frac{\partial v}{\partial t} + \frac{3}{2\epsilon^2} \frac{\partial v}{\partial x} \right) = \left(\frac{\epsilon}{\sqrt{6}} v + \frac{\sqrt{6}}{2\epsilon} \right)^2 \frac{\epsilon}{\sqrt{6}} \frac{\partial v}{\partial x} + \frac{\epsilon}{\sqrt{6}} \frac{\partial^3 v}{\partial x^3}$$

$$\text{or,} \quad \frac{\partial v}{\partial t} + \frac{3}{2\epsilon^2} \frac{\partial v}{\partial x} = \left(\frac{\epsilon^2}{6} v^2 + v + \frac{3}{2\epsilon^2} \right) \frac{\partial v}{\partial x} + \frac{\partial^3 v}{\partial x^3}$$

$$\text{or,} \quad \frac{\partial v}{\partial t} = \left(\frac{\epsilon^2}{6} v^2 + v \right) \frac{\partial v}{\partial x} + \frac{\partial^3 v}{\partial x^3} \qquad\qquad (3.20)$$

Thus we see that the MKdV equation is not Galilean invariant. But we also recognize that this transformed equation interpolates between the KdV and the MKdV equations in the sense that when $\epsilon=0$, Eq. (3.20) reduces to

$$\frac{\partial v}{\partial t} = v \frac{\partial v}{\partial x} + \frac{\partial^3 v}{\partial x^3}$$

which is the KdV equation with $v(x,t)$ as the dynamical variable. On the other hand, if $\epsilon \to \infty$, then under a rescaling $v \to \frac{\epsilon}{\sqrt{6}} v$, the equation becomes

$$\frac{\partial v}{\partial t} = v^2 \frac{\partial v}{\partial x} + \frac{\partial^3 v}{\partial x^3}$$

which is the MKdV equation of Eq. (3.16). We could have seen this also from the fact that the Riccati relation of Eq. (3.17)

$$u = v^2 + i\sqrt{6} \frac{\partial v}{\partial x}$$

goes over, under the Galilean transformation of Eq. (3.19), to

$$u + \frac{3}{2\epsilon^2} = \left(\frac{\epsilon^2}{6} v^2 + v + \frac{3}{2\epsilon^2}\right) + i\sqrt{6} \frac{\epsilon}{\sqrt{6}} \frac{\partial v}{\partial x}$$

$$\text{or,} \quad u = \frac{\epsilon^2}{6} v^2 + v + i\epsilon \frac{\partial v}{\partial x} \qquad (3.21)$$

For $\epsilon=0$, this implies the KdV equation since

$$u = v$$

For $\epsilon \to \infty$, on the other hand, with $v \to \dfrac{\epsilon}{\sqrt{6}} \, v$ we obtain the MKdV equation since

$$u = v^2 + i\sqrt{6} \, \frac{\partial v}{\partial x} \qquad\qquad (3.22)$$

Incidentally, one should not be worried about the occurrence of imaginary factors. This, simply, is a consequence of our particular choice of the coefficients in the KdV and the MKdV equations.

Infinite Number of Conserved Quantities:

We are now in a position to prove that the KdV equation possesses an infinite number of conserved quantities. Let us note that if $v(x,t)$ is a solution of the generalized equation

$$\frac{\partial v}{\partial t} = \left(\frac{\epsilon^2}{6} \, v^2 + v\right) \frac{\partial v}{\partial x} + \frac{\partial^3 v}{\partial x^3}$$

$$= \frac{\partial}{\partial x} \left(\frac{\epsilon^2}{18} \, v^3 + \frac{1}{2} \, v^2 + \frac{\partial^2 v}{\partial x^2}\right) \qquad\qquad (3.23)$$

then it also gives a solution of the KdV equation through the relation

$$u(x,t) = \frac{\epsilon^2}{6} v^2 + v + i\epsilon \frac{\partial v}{\partial x} \tag{3.24}$$

Furthermore, we also recognize that the generalized equation, Eq. (3.23), is already in the form of a continuity equation so that

$$\frac{dK}{dt} = \frac{d}{dt} \int_{-\infty}^{\infty} dx \, v(x,t) = \int_{-\infty}^{\infty} dx \, \frac{\partial v(x,t)}{\partial t} = 0 \tag{3.25}$$

Therefore, we can identify

$$K = \int_{-\infty}^{\infty} dx \, \rho[v(x,t)]$$

where $\qquad\qquad\qquad\qquad\qquad\qquad\qquad\qquad\qquad$ (3.26)

$$\rho[v(x,t)] = v(x,t)$$

Note that $u(x,t)$ is related to $v(x,t)$ through the generalized nonlinear relation given in Eq. (3.24). We can invert this, formally, to expand $v(x,t)$ in terms of $u(x,t)$ as

$$v(x,t) = \sum_{n=0}^{\infty} \epsilon^n v_n[u(x,t)] \tag{3.27}$$

From Eqs. (3.23) and (3.26), we see that this would give the $v_n[u(x,t)]$'s as the conserved densities for the KdV equation since each power of ϵ must independently satisfy a continuity equation.

Let us, however, note that a conserved density must not be a total derivative - otherwise, it would lead to a trivial conserved quantity. Thus we must make sure that this expansion of v in terms of u's is not trivial - i.e., it is not a total derivative. This can be seen as follows. Since we see from Eq. (3.24) that

$$u = v + i\epsilon \frac{\partial v}{\partial x} + \frac{\epsilon^2}{6} v^2 \qquad (3.28)$$

we can ask whether the inversion of this relation would lead to pure polynomial terms in u's because such terms cannot be written as total derivatives. Note that even if v is a pure polynomial in u's, $\frac{\partial v}{\partial x}$ would involve $\frac{\partial u}{\partial x}$ terms and hence to investigate such a question we can safely ignore the second term on the right hand side of Eq. (3.28). Thus, our question is whether

$$u = v + \frac{\epsilon^2}{6} v^2 \qquad (3.29)$$

allows for v to be expressible completely in terms of polynomials in u's. Note that if

$$u = v + \frac{\epsilon^2}{6} v^2$$

then

$$u + \frac{6}{4\epsilon^2} = \left(\frac{\sqrt{6}}{2\epsilon} + \frac{\epsilon}{\sqrt{6}} v\right)^2$$

or, $\quad \dfrac{\sqrt{6}}{2\epsilon} + \dfrac{\epsilon}{\sqrt{6}} \; v = \left(u + \dfrac{6}{4\epsilon^2} \right)^{1/2}$

or, $\quad \dfrac{\epsilon}{\sqrt{6}} \; v = \left(u + \dfrac{6}{4\epsilon^2} \right)^{1/2} - \dfrac{\sqrt{6}}{2\epsilon}$

or, $\quad v = \dfrac{\sqrt{6}}{\epsilon} \left[\left(u + \dfrac{3}{2\epsilon^2} \right)^{1/2} - \dfrac{\sqrt{6}}{2\epsilon} \right]$

or, $\quad v = \dfrac{3}{\epsilon^2} \left[\left(1 + \dfrac{2}{3} \, \epsilon^2 u \right)^{1/2} - 1 \right]$ \hfill (3.30)

Thus we see that v indeed possesses pure polynomial terms in u's and furthermore, these are of even power in ϵ.

Next let us show that the odd powers of ϵ in the expansion of v are total derivatives and hence are trivial. Let us note from Eq. (3.28) that v is in general complex although u is real. Thus let us decompose

$$v = y + iz \hspace{4cm} (3.31)$$

where both y and z are real. Now, since

$$u = v + i\epsilon \, \frac{\partial v}{\partial x} + \frac{\epsilon^2}{6} \, v^2$$

in terms of the new variables, it has the form

$$u = y + iz + i\epsilon \, \frac{\partial}{\partial x} \, (y+iz) + \frac{\epsilon^2}{6} \, (y^2 - z^2 + 2iyz)$$

or, $u = \left(y - \epsilon \frac{\partial z}{\partial x} + \frac{\epsilon^2}{6} (y^2 - z^2)\right)$

$$+ i\left(z + \epsilon \frac{\partial y}{\partial x} + \frac{\epsilon^2}{3} yz\right) \qquad (3.32)$$

But since u is real, we must have

$$z + \epsilon \frac{\partial y}{\partial x} + \frac{\epsilon^2}{3} yz = 0$$

or, $z\left(1 + \frac{\epsilon^2}{3} y\right) = - \epsilon \frac{\partial y}{\partial x}$

or, $z = -\epsilon \frac{\partial y}{\partial x} / \left(1 + \frac{\epsilon^2}{3} y\right) = - \frac{3}{\epsilon} \frac{\partial}{\partial x} \ell n\left(1 + \frac{\epsilon^2}{3} y\right)$ (3.33)

This shows that the imaginary part of v is a pure derivative and involves terms which have explicitly odd power in ϵ.

Although this argument shows that the imaginary part of v is of odd powers in ϵ and is a total derivative, it is not clear that there are no odd power ϵ terms in the real part of v. That this is indeed the case can be understood from the scaling properties of various quantities.

We know the scaling behavior of x and u from Eq. (1.40). Consequently, we can determine the scaling behavior of ϵ and v from Eq. (3.28). This gives the scaling dimensions

$$[u] = 1$$
$$[v] = 1 \qquad\qquad (3.34)$$
$$[x] = - \frac{1}{2}$$
$$[\epsilon] = - \frac{1}{2}$$

Thus it is clear that any term that has an odd power of ϵ's must necessarily have an odd number of derivatives also since v has scaling dimension of an integer. But a derivative term necessarily carries a factor of i as is seen from Eq. (3.28) and hence such a term must be imaginary. By the same argument, the real terms must all be of even powers of ϵ. And our analysis above shows that it is the real part, which is even in powers of ϵ, which gives rise to nontrivial conserved quantities.

We are now ready to construct the conserved quantities. Substituting the expansion of v[u] from Eq. (3.27) into the definition of u in Eq. (3.28), we obtain

$$u = v + i \epsilon \frac{\partial v}{\partial x} + \frac{\epsilon^2}{6} v^2$$

$$= \sum_{n=0}^{\infty} \epsilon^n v_n + i \sum_{n=0}^{\infty} \epsilon^{n+1} \frac{\partial v_n}{\partial x} + \sum_{n=0}^{\infty} \frac{\epsilon^{n+2}}{6} \sum_{m=0}^{n} v_{n-m} v_m$$

$$\text{or, } u = \sum_{n=0}^{\infty} \epsilon^n \left[v_n + i \frac{\partial v_{n-1}}{\partial x} + \frac{1}{6} \sum_{m=0}^{n-2} v_{n-m-2} v_m \right] \qquad (3.35)$$

where we have assumed

$$v_{-1} = 0 = v_{-2}$$

Furthermore, since u is independent of ϵ, comparing ϵ^0 terms, we obtain

$$u = v_0 \qquad\qquad (3.36)$$

Equating the coefficients of the ϵ-dependent terms in Eq. (3.35), we obtain

$$v_n + i\,\frac{\partial v_{n-1}}{\partial x} + \frac{1}{6}\sum_{m=0}^{n-2} v_{n-m-2}v_m = 0 \qquad n > 0 \qquad (3.37)$$

This gives a recursion relation between the various conserved densities which in turn allows their construction. For example,

$$v_1 + i\,\frac{\partial v_0}{\partial x} = 0$$

$$\text{or,} \quad v_1 = -i\,\frac{\partial u}{\partial x} \qquad\qquad (3.38)$$

$$v_2 + i\,\frac{\partial v_1}{\partial x} + \frac{1}{6}v_0^2 = 0$$

$$\text{or,} \quad v_2 = -\,\frac{\partial^2 u}{\partial x^2} - \frac{1}{6}u^2 = -\frac{1}{6}u^2 - \frac{\partial^2 u}{\partial x^2} \qquad\qquad (3.39)$$

$$v_3 + i\,\frac{\partial v_2}{\partial x} + \frac{1}{6}(v_1 v_0 + v_0 v_1) = 0$$

$$\text{or,} \quad v_3 = -i\,\frac{\partial v_2}{\partial x} - \frac{1}{3}v_1 v_0$$

$$= -i\,\frac{\partial}{\partial x}\left(-\frac{1}{6}u^2 - \frac{\partial^2 u}{\partial x^2}\right) - \frac{1}{3}(-i)\,\frac{\partial u}{\partial x}\,u$$

$$= -i\,\frac{\partial}{\partial x}\left(-\frac{1}{6}u^2 - \frac{\partial^2 u}{\partial x^2}\right) + \frac{i}{6}\,\frac{\partial}{\partial x}(u^2)$$

$$\text{or, } v_3 = i \frac{\partial}{\partial x} \left(\frac{1}{3} u^2 + \frac{\partial^2 u}{\partial x^2} \right) \tag{3.40}$$

$$v_4 + i \frac{\partial v_3}{\partial x} + \frac{1}{6} \left(v_2 v_0 + v_1^2 + v_0 v_2 \right) = 0$$

$$\text{or, } v_4 = -i \frac{\partial v_3}{\partial x} - \frac{1}{6} \left(2 v_2 v_0 + v_1^2 \right)$$

$$= \frac{\partial^2}{\partial x^2} \left(\frac{1}{3} u^2 + \frac{\partial^2 u}{\partial x^2} \right) - \frac{1}{6} \left[2 \left(- \frac{1}{6} u^2 - \frac{\partial^2 u}{\partial x^2} \right) u - \left(\frac{\partial u}{\partial x} \right)^2 \right]$$

$$= \frac{\partial^2}{\partial x^2} \left(\frac{1}{3} u^2 + \frac{\partial^2 u}{\partial x^2} \right) + \frac{1}{18} u^3 + \frac{1}{3} \frac{\partial}{\partial x} \left(u \frac{\partial u}{\partial x} \right)$$

$$- \frac{1}{3} \left(\frac{\partial u}{\partial x} \right)^2 + \frac{1}{6} \left(\frac{\partial u}{\partial x} \right)^2$$

$$= \frac{\partial^2}{\partial x^2} \left(\frac{1}{3} u^2 + \frac{\partial^2 u}{\partial x^2} + \frac{1}{6} u^2 \right) + \frac{1}{18} u^3 - \frac{1}{6} \left(\frac{\partial u}{\partial x} \right)^2$$

$$\text{or, } v_4 = \frac{1}{3} \left(\frac{1}{6} u^3 - \frac{1}{2} \left(\frac{\partial u}{\partial x} \right)^2 \right) + \frac{\partial^2}{\partial x^2} \left(\frac{1}{2} u^2 + \frac{\partial^2 u}{\partial x^2} \right) \tag{3.41}$$

It is clear that we can construct the conserved densities recursively. There are several things worth noting here. First of all, as we had noted earlier, the odd powers of ϵ do give rise to imaginary terms which are total derivatives. The real terms which are coefficients of the even powers of ϵ do contain pure monomials of u and hence are not total derivatives. More importantly let us recall that the conserved densities can be arbitrary up to multiplicative constants and addition of total derivatives. If we keep these things in mind, it becomes clear that

$$\rho_0 = 3v_0 = 3u$$

$$\rho_1 = -3v_2 = \frac{1}{2} u^2 + \text{total derivatives} \tag{3.42}$$

$$\rho_2 = 3v_4 = \frac{1}{6} u^3 - \frac{1}{2} (\frac{\partial u}{\partial x})^2 + \text{total derivatives}$$

Comparing with Eqs. (3.7), (3.9) and (3.12), we see that the conserved quantities we had obtained earlier are simply the first three nontrivial densities in the power series expansion in ϵ given in Eq. (3.27). The nth conserved density is given as

$$\rho_n = 3(-1)^n v_{2n} \tag{3.43}$$

so that

$$H_n = \int_{-\infty}^{\infty} dx \, \rho_n = 3(-1)^n \int_{-\infty}^{\infty} dx \, v_{2n} \tag{3.44}$$

Let us note that the scaling behavior of each of the v_n's can be calculated as follows. Since

$$[\frac{\partial}{\partial x}] = \frac{1}{2}$$

therefore, from Eq. (3.37) we see that

$$[v_n] = [v_{n-1}] + \frac{1}{2} \qquad\qquad\qquad (3.45)$$

Given $[v_0] = [u] = 1$, it then follows that

$$[v_n] = \frac{n}{2} + 1 \qquad\qquad\qquad (3.46)$$

Since the nontrivial conserved densities are identified as

$$\rho_n \sim v_{2n}$$

therefore,

$$[\rho_n] = [v_{2n}] = n+1 \qquad\qquad\qquad (3.47)$$

This shows that there is a denumerably infinite number of conserved quantities each scaling with a distinct integer power law. The uniqueness of one conserved quantity for each integer valued scaling law takes a little while to prove but it has been shown that there exists exactly one such conserved quantity given a scaling behavior. Once again uniqueness only holds up to a multiplicative constant and addition of total derivative terms.

Secondly, since each conserved quantity has a distinct scaling behavior, it is clear that they are independent.

Integrability of the KdV system:

So far we have shown two of the three requirements needed to exhibit integrability of the KdV system. Namely, we have shown that there exists an infinite number of conserved quantities and that they are independent. What remains to be shown is that these quantities are in involution.

To show that, let us note from Eqs. (3.7), (3.10) and (3.12) that the conserved quantities

$$H_0 = 3 \int_{-\infty}^{\infty} dx \; u(x,t)$$

$$H_1 = \int_{-\infty}^{\infty} dx \; \frac{1}{2} u^2(x,t) \tag{3.48}$$

$$H_2 = \int_{-\infty}^{\infty} dx \left(\frac{1}{3!} u^3 - \frac{1}{2} \left(\frac{\partial u}{\partial x} \right)^2 \right)$$

satisfy the functional relation

$$\left(D^3 + \frac{1}{3} (Du+uD) \right) \frac{\delta H_{n-1}}{\delta u(x)} = D \frac{\delta H_n}{\delta u(x)} \qquad n = 0,1,2 \tag{3.49}$$

This can be checked as follows. The relation above clearly holds for n=0 if we assume $H_{-1} = 0$. For n=1, Eq. (3.49) becomes

$$\left(D^3 + \frac{1}{3} (Du+uD) \right) \frac{\delta H_0}{\delta u(x)} = D \frac{\delta H_1}{\delta u(x)}$$

$$\text{L.H.S.} = \int_{-\infty}^{\infty} dy \left(D_x^3 + \frac{1}{3} (D_x u(x) + u(x) D_x) \right) 3\delta(x-y)$$

$$= \int_{-\infty}^{\infty} dy \left[(D_x u(x)) \delta(x-y) + \frac{2}{3} 3 u(x) D_x \delta(x-y) \right]$$

$$= D_x u(x)$$

and

$$\text{R.H.S.} = D \frac{\delta H_1}{\delta u(x)} = D_x u(x)$$

so that

$$\text{L.H.S.} = \text{R.H.S.}$$

For n=2

$$(D^3 + \frac{1}{3} (Du+uD)) \frac{\delta H_1}{\delta u(x)} = D \frac{\delta H_2}{\delta u(x)}$$

$$\text{L.H.S.} = \left(D_x^3 + \frac{1}{3} (D_x u(x) + u(x) D_x) \right) u(x)$$

$$= D_x^3 u(x) + u(x) D_x u(x)$$

whereas

$$\text{R.H.S.} = D \frac{\delta H_2}{\delta u(x)} = D_x \left(\frac{1}{2} u^2(x) + D_x^2 u(x) \right)$$

$$= u(x) D_x u(x) + D_x^3 u(x)$$

so that

$$L.H.S. = R.H.S.$$

This suggests that all the conserved quantities may obey a similar functional recursion relation. Namely, that one can choose the Hamiltonians such that

$$\left(D_x^3 + \frac{1}{3} \left(D_x u(x) + u(x) D_x\right)\right) \frac{\delta H_{n-1}}{\delta u(x)} = D_x \frac{\delta H_n}{\delta u(x)} \qquad \forall\ n \qquad (3.50)$$

That this can be done can be proved inductively as follows. Suppose that the recursion relation of Eq. (3.50) holds for $n = 1, 2, \ldots m$ where m is a fixed number. Thus the identity when $n=m$ is

$$\left(D_x^3 + \frac{1}{3} \left(D_x u(x) + u(x) D_x\right)\right) \frac{\delta H_{m-1}}{\delta u(x)} = D_x \frac{\delta H_m}{\delta u(x)} \qquad (3.51)$$

But since H_m is a conserved quantity

$$\frac{dH_m}{dt} = 0 \qquad (3.52)$$

This time evolution can be calculated as

$$\frac{dH_m}{dt} = \{H_m, H_2\}_1$$

where

$$\{u(x),u(y)\}_1 = D_x \delta(x-y) \qquad (3.53)$$

We can also calculate this using the second Poisson bracket relation of Eq. (1.63), namely,

$$\frac{dH_m}{dt} = \{H_m,H_1\}_2$$

where

$$\{u(x),u(y)\}_2 = \left(D_x^3 + \frac{1}{3}\,(D_x u(x)+u(x)D_x)\right)\delta(x-y) \qquad (3.54)$$

Using the form in Eq. (3.54), we obtain

$$\frac{dH_m}{dt} = \{H_m,H_1\}_2$$

$$= \int_{-\infty}^{\infty} dxdy\, \frac{\delta H_m}{\delta u(x)}\,\left(D_x^3 + \frac{1}{3}\,(D_x u(x)+u(x)D_x)\right)\,\delta(x-y)\,\frac{\delta H_1}{\delta u(y)}$$

$$= -\int_{-\infty}^{\infty} dx\left((D_x^3 + \frac{1}{3}\,(D_x u(x)+u(x)D_x))\right)\frac{\delta H_m}{\delta u(x)}\,u(x) \qquad (3.55)$$

Since H_m is conserved, this must vanish, and consequently, the integrand in Eq. (3.55) must be a total derivative. This is possible only if

$$\left(D_x^3 + \frac{1}{3}\, (D_x u(x) + u(x) D_x)\right) \frac{\delta H_m}{\delta u(x)} = D_x \frac{\delta K_m}{\delta u(x)} \qquad (3.56)$$

Counting the scaling behavior we see that K_m should scale one power higher than H_m. Furthermore, we can also show that K_m is conserved. But since there is only one conserved quantity with a specific scaling behavior, we can identify

$$K_m = H_{m+1} \qquad (3.57)$$

so that

$$\left(D_x^3 + \frac{1}{3}\, (D_x u(x) + u(x) D_x)\right) \frac{\delta H_m}{\delta u(x)} = D_x \frac{\delta H_{m+1}}{\delta u(x)} \qquad (3.58)$$

Thus we have shown that if the recursion relation of Eq. (3.50) is valid for $n = 1,2,3,\ldots m$, then it is valid for $n = m+1$ also. By induction then it holds for all n. This recursion relation is crucial in proving that the conserved quantities are in involution. In fact, since

$$\left(D_x^3 + \frac{1}{3}\, (D_x u(x) + u(x) D_x)\right) \frac{\delta H_{n-1}}{\delta u(x)} = D_x \frac{\delta H_n}{\delta u(x)} \qquad \forall\ n$$

therefore,

$$\{H_n, H_m\}_1 = \int_{-\infty}^{\infty} dx\ \frac{\delta H_n}{\delta u(x)}\ D_x\ \frac{\delta H_m}{\delta u(x)}$$

$$= -\int_{-\infty}^{\infty} dx\ D_x\ \frac{\delta H_n}{\delta u(x)}\ \frac{\delta H_m}{\delta u(x)}$$

$$= - \int_{-\infty}^{\infty} dx \left(D_x^3 + \frac{1}{3} \left(D_x u(x) + u(x) D_x \right) \right) \frac{\delta H_{n-1}}{\delta u(x)} \frac{\delta H_m}{\delta u(x)}$$

$$= \int_{-\infty}^{\infty} dx \frac{\delta H_{n-1}}{\delta u(x)} \left(D_x^3 + \frac{1}{3} \left(D_x u(x) + u(x) D_x \right) \right) \frac{\delta H_m}{\delta u(x)}$$

$$= \int_{-\infty}^{\infty} dx \frac{\delta H_{n-1}}{\delta u(x)} D_x \frac{\delta H_{m+1}}{\delta u(x)}$$

$$= \{ H_{n-1}, H_{m+1} \}_1 \qquad\qquad (3.59)$$

By iteration we can then show that

$$\{ H_n, H_m \}_1 = \{ H_m, H_n \}_1 = 0 \qquad\qquad (3.60)$$

This proves that all the conserved quantities are not only independent but they are also in involution. One can also show in a similar manner, using Eq. (3.50), that

$$\{ H_n, H_m \}_2 = 0 \qquad\qquad (3.61)$$

That all these H_n's are conserved follows from the relation

$$\frac{dH_n}{dt} = \{ H_n, H_2 \}_1 = 0 = \{ H_n, H_1 \}_2 \qquad\qquad (3.62)$$

This proves, by Liouville's theorem, that the KdV equation is integrable. Incidentally since the MKdV equation also shares the same conserved quantities, this also proves that the MKdV equation is integrable.

Higher Order Equations of the Hierarchy:

We have seen that the KdV equation has an infinite number of conserved quantities H_n, $n = 0,1,2,\ldots\infty$, such that

$$\{H_n, H_m\}_1 = \{H_n, H_m\}_2 = 0$$

Namely, they are in involution with respect to either of the Poisson bracket structures associated with the KdV equation. We also saw that the conserved quantities satisfy the functional recursion relation

$$(D^3 + \frac{1}{3}(Du+uD)) \frac{\delta H_n}{\delta u(x)} = D \frac{\delta H_{n+1}}{\delta u(x)}$$

This explains why the KdV equation can be written in two different ways, namely,

$$\frac{\partial u}{\partial t} = \{u(x), H_1\}_2 = (D^3 + \frac{1}{3}(Du+uD)) \frac{\delta H_1}{\delta u(x)}$$

$$= D \frac{\delta H_2}{\delta u(x)} = \{u(x), H_2\}_1 \qquad (3.63)$$

Each of the conserved quantities, of course, can be thought of as a Hamiltonian and it generates its own evolution equation given by

$$\frac{\partial u}{\partial t} = \{u(x), H_n\}_2 = (D^3 + \frac{1}{3}(Du+uD)) \frac{\delta H_n}{\delta u(x)}$$

$$= D \frac{\delta H_{n+1}}{\delta u(x)} = \{u(x), H_{n+1}\}_1 \qquad (3.64)$$

These are known as the higher order equations of the KdV hierarchy. In fact, it is clear that every integrable system must possess a hierarchial structure of evolution equations. Furthermore, since the H_n's are in involution, each equation in the hierarchy shares the same conserved quantities and is integrable. It is, therefore, instructive to think of u's as a function of an infinite number of time variables. Thus

$$u = u(x, t_0, t_1, t_2 \ldots) \qquad (3.65)$$

where the t_n's represent evolution parameters with respect to the Hamiltonian H_n. Thus for a given evolution equation of the hierarchy, only the corresponding time parameter changes and the others stay constant. Furthermore, since the H_n's are in involution, the different flows are easily seen to commute. Namely, if we evolve the system along t_n for a time Δt_n and then along t_m for a time Δt_m, the final state would be the same as if we had evolved the system along t_m for a time Δt_m first and then along t_n for a time Δt_n.

Let us, for completeness, note from Eq. (3.64) that the first equation of the KdV hierarchy is

$$\frac{\partial u}{\partial t} = D \; \frac{\delta H_1}{\delta u(x)} = \frac{\partial u}{\partial x} \tag{3.66}$$

This is nothing other than the equation for a chiral particle or a chiral wave in the sense that it describes a wave moving only to the left. This is a characteristic of the KdV hierarchy as we have seen in Eq. (2.24), in the case of the KdV equation, namely, their solutions are chiral.

References:

Das, A., Phys. Lett. B207, 429 (1988).

Gardner, C. S., J. M. Greene, M. D. Kruskal and R. M. Miura, Comm. Pure Appl. Math. 27, 97 (1974).

Kruskal, M. D., R. M. Miura, C. S. Gardner and N. Zabusky, J. Math. Phys. 11, 952 (1970).

Magri, F., J. Math. Phys. 19, 1156 (1978).

Miura, R. M., J. Math. Phys. 9, 1202 (1968).

Miura, R. M., C. S. Gardner and M. D. Kruskal, J. Math. Phys. 9, 1204 (1968).

CHAPTER 4

INITIAL VALUE PROBLEM FOR THE KDV EQUATION

There exist well known methods for solving a given linear Hamiltonian system with fixed initial conditions. For example, the Laplace transformation makes a partial differential equation into an ordinary one which is then solved. Similarly, the Fourier transformation changes a differential equation into an algebraic one which is easily solved. These methods are, however, inapplicable to a nonlinear system.

Gardner, Greene, Kruskal and Miura were the first to solve the initial value problem for the KdV equation in a very ingenious way. In the subsequent years this method has been put into firm footing and has become the standard method for solving nonlinear systems. This goes by the name of inverse scattering theory which we shall study in detail. But let us first analyze the Gardner-Greene-Kruskal-Miura solution of the KdV equation.

The Schrödinger Equation:

Let us consider the time independent Schrödinger equation described by

$$\frac{\partial^2 \psi}{\partial x^2} + (\frac{1}{6} u(x,t) + \lambda)\psi = 0 \tag{4.1}$$

where u(x,t) is the KdV variable which satisfies

$$\frac{\partial u}{\partial t} = u \frac{\partial u}{\partial x} + \frac{\partial^3 u}{\partial x^3} \qquad\qquad (4.2)$$

The variable t in u(x,t) is not the time variable for the Schrödinger equation. Rather it should be considered as a parameter characterizing the potential in the Schrödinger equation. Furthermore, λ is the energy eigenvalue and ψ the corresponding eigenfunction both of which will in general depend on t.

The origin and the relevance of the Schrödinger equation in the study of the KdV equation is mysterious at this point and we will discuss this in more detail later. But for the moment, let us note that we can use the Schrödinger equation, Eq. (4.1), to eliminate u partially in favor of the wave function ψ in the KdV equation. For example, from Eq. (4.1) we see that

$$u(x,t) = -6 \left(\lambda + \frac{\partial^2 \psi}{\partial x^2} / \psi \right)$$

$$= -6 \left(\lambda + \frac{\psi_{xx}}{\psi} \right) \qquad\qquad (4.3)$$

From here on we use the convention that subscripts t and x denote differentiation with respect to those variables. Using Eq. (4.3), we can easily calculate

$$u_t = -6 \left(\lambda_t + \frac{\psi_{xxt}}{\psi} - \frac{\psi_{xx}\psi_t}{\psi^2} \right)$$

$$= -\frac{6}{\psi^2}\left(\lambda_t\psi^2 + (\psi\psi_{xxt}-\psi_{xx}\psi_t)\right)$$

$$= -\frac{6}{\psi^2}\left(\lambda_t\psi^2 + \frac{\partial}{\partial x}\psi^2\frac{\partial}{\partial x}\left(\frac{\psi_t}{\psi}\right)\right) \tag{4.4}$$

$$u_x = -6\left(\frac{\psi_{xxx}}{\psi} - \frac{\psi_{xx}\psi_x}{\psi^2}\right)$$

$$= -\frac{6}{\psi^2}\left(\psi\psi_{xxx}-\psi_{xx}\psi_x\right)$$

$$= -\frac{6}{\psi^2}\frac{\partial}{\partial x}\psi^2\frac{\partial}{\partial x}\left(\frac{\psi_x}{\psi}\right) \tag{4.5}$$

$$uu_x = -\frac{6u}{\psi^2}\frac{\partial}{\partial x}\psi^2\frac{\partial}{\partial x}\left(\frac{\psi_x}{\psi}\right)$$

$$= -\frac{6}{\psi^2}\frac{\partial}{\partial x}u\psi^2\frac{\partial}{\partial x}\left(\frac{\psi_x}{\psi}\right) + 6u_x\frac{\partial}{\partial x}\left(\frac{\psi_x}{\psi}\right)$$

$$= -\frac{6}{\psi^2}\frac{\partial}{\partial x}\psi^2\frac{\partial}{\partial x}\left(\frac{u\psi_x}{\psi}\right) + \frac{6}{\psi^2}\frac{\partial}{\partial x}\left(\psi\psi_x u_x\right)$$

$$+ 6u_x\left(\frac{\psi_{xx}}{\psi} - \left(\frac{\psi_x}{\psi}\right)^2\right)$$

$$= -\frac{6}{\psi^2}\frac{\partial}{\partial x}\psi^2\frac{\partial}{\partial x}\left(\frac{u\psi_x}{\psi}\right) + \frac{6}{\psi^2}\left(\psi_x^2 u_x+\psi\psi_{xx}u_x+\psi\psi_x u_{xx}\right)$$

$$+ 6u_x\left(\frac{\psi_{xx}}{\psi} - \left(\frac{\psi_x}{\psi}\right)^2\right)$$

$$= - \frac{6}{\psi^2} \frac{\partial}{\partial x} \psi^2 \frac{\partial}{\partial x} \left(\frac{u\psi_x}{\psi} \right) + \frac{12u_x \psi_{xx}}{\psi} + \frac{6u_{xx}\psi_x}{\psi}$$

or, $$uu_x = - \frac{6}{\psi^2} \frac{\partial}{\partial x} \psi^2 \left(\frac{u\psi_x}{\psi} \right) + 12u_x \left(- \frac{1}{6} u - \lambda \right) + \frac{6u_{xx}\psi_x}{\psi}$$

or, $$3uu_x = - \frac{6}{\psi^2} \frac{\partial}{\partial x} \psi^2 \frac{\partial}{\partial x} \left(\frac{u\psi_x}{\psi} \right) - 12\lambda u_x + \frac{6u_{xx}\psi_x}{\psi}$$

Using the formula for u_x we then obtain

$$uu_x = - \frac{6}{\psi^2} \frac{\partial}{\partial x} \psi^2 \frac{\partial}{\partial x} \left(\left(\frac{1}{3} u - 4\lambda \right) \frac{\psi_x}{\psi} \right) + \frac{2u_{xx}\psi_x}{\psi} \quad (4.6)$$

Similarly, we can calculate

$$u_{xxx} = \frac{\partial^2}{\partial x^2} \left(-6 \left(\frac{\psi_{xxx}}{\psi} - \frac{\psi_{xx}\psi_x}{\psi^2} \right) \right)$$

$$= -6 \frac{\partial^2}{\partial x^2} \left(\frac{\psi_{xxx}}{\psi} \right) + 6 \frac{\partial^2}{\partial x^2} \left(\frac{\psi_{xx}\psi_x}{\psi^2} \right)$$

$$= - \frac{6}{\psi^2} \frac{\partial}{\partial x} \psi^2 \frac{\partial}{\partial x} \left(\frac{\psi_{xxx}}{\psi} \right) + 12 \frac{\psi_x}{\psi} \frac{\partial}{\partial x} \left(\frac{\psi_{xxx}}{\psi} \right)$$

$$+ \frac{6}{\psi^2} \frac{\partial}{\partial x} \psi^2 \frac{\partial}{\partial x} \left(\frac{\psi_{xx}\psi_x}{\psi^2} \right) - 12 \frac{\psi_x}{\psi} \frac{\partial}{\partial x} \left(\frac{\psi_{xx}\psi_x}{\psi^2} \right)$$

$$= - \frac{6}{\psi^2} \frac{\partial}{\partial x} \psi^2 \frac{\partial}{\partial x} \left(\frac{\psi_{xxx}}{\psi} - \frac{\psi_{xx}}{\psi} \frac{\psi_x}{\psi} \right)$$

$$+ 12 \frac{\psi_x}{\psi} \frac{\partial}{\partial x} \left(\frac{\psi_{xxx}}{\psi} - \frac{\psi_{xx}\psi_x}{\psi^2} \right)$$

$$= -\frac{6}{\psi^2} \frac{\partial}{\partial x} \psi^2 \frac{\partial}{\partial x} \left(\frac{\psi_{xxx}}{\psi} + (\frac{1}{6} u + \lambda) \frac{\psi_x}{\psi}\right)$$

$$+ 12 \frac{\psi_x}{\psi} \frac{\partial^2}{\partial x^2} \left(\frac{\psi_{xx}}{\psi}\right)$$

$$= -\frac{6}{\psi^2} \frac{\partial}{\partial x} \psi^2 \frac{\partial}{\partial x} \left(\frac{\psi_{xxx}}{\psi} + (\frac{1}{6} u + \lambda) \frac{\psi_x}{\psi}\right)$$

$$+ 12 \frac{\psi_x}{\psi} \frac{\partial^2}{\partial x^2} \left(-\frac{1}{6} u\right)$$

$$\text{or,} \quad u_{xxx} = -\frac{6}{\psi^2} \frac{\partial}{\partial x} \psi^2 \frac{\partial}{\partial x} \left(\frac{\psi_{xxx}}{\psi} + (\frac{1}{6} u + \lambda) \frac{\psi_x}{\psi}\right)$$

$$- 2 \frac{\psi_x u_{xx}}{\psi} \tag{4.7}$$

Thus the KdV equation can be obtained in terms of the Schrödinger wave function, from Eqs. (4.4), (4.6) and (4.7), to be

$$u_t - u u_x - u_{xxx} = 0$$

$$\text{or,} \quad -\frac{6}{\psi^2} \left[\frac{\partial}{\partial x} \psi^2 \frac{\partial}{\partial x} \left(\frac{\psi_t}{\psi} - (\frac{1}{3} u - 4\lambda) \frac{\psi_x}{\psi} - \frac{\psi_{xxx}}{\psi}\right.\right.$$

$$\left.\left. - (\frac{1}{6} u + \lambda) \frac{\psi_x}{\psi}\right) + \lambda_t \psi^2\right] = 0$$

$$\text{or,} \quad \lambda_t \psi^2 - \frac{\partial}{\partial x} \psi^2 \frac{\partial}{\partial x} \left(\frac{(\psi_{xxx} - \psi_t + (\frac{1}{2} u - 3\lambda)\psi_x)}{\psi}\right) = 0 \tag{4.8}$$

Integrating this over x and remembering that the wave function ψ vanishes at x = $\pm\infty$ (both for the bound state as well as for the oscillatory states) we obtain

$$\lambda_t = 0 \qquad\qquad (4.9)$$

This is an important conclusion.   Namely, if u(x,t) evolves according to the KdV equation, then the eigenvalues of the Schrödinger equation, Eq. (4.1), with   u(x,t) as the potential are independent of the parameter t.   Since λ_t vanishes, we see Eq. (4.8) to give

$$\frac{\partial}{\partial x} \left(\frac{(\psi_{xxx} - \psi_t + (\frac{1}{2} u - 3\lambda)\psi_x)}{\psi} \right) = 0 \qquad\qquad (4.10)$$

Using the Schrödinger equation (Eq. (4.1)) this can be written as

$$\frac{\partial}{\partial x} \left(\frac{(\psi_t + \frac{1}{6} u_x \psi + 4\lambda\psi_x - \frac{1}{3} u\psi_x)}{\psi} \right) = 0$$

It follows, therefore, that

$$\frac{\psi_t + \frac{1}{6} u_x \psi + 4\lambda\psi_x - \frac{1}{3} u\psi_x}{\psi} = \text{constant} \qquad\qquad (4.11)$$

Since the constant is independent of x, it can be determined from the asymptotic behavior of various quantities.

Time Evolution of Scattering Parameters:

   i)   Bound States: $\lambda = -\kappa^2 < 0$.

   Let us recall that since

$$\lambda_t = 0$$

evolution in the parameter t must represent a unitary symmetry of the Schrödinger system. Consequently, one could study the system at $t=0$. In this case, we know that the bound state wave functions fall off exponentially so that

$$\psi(x) \xrightarrow[x \to \infty]{} e^{-\kappa x} \tag{4.12}$$

Furthermore, we know that the KdV variable satisfies

$$u(x,t) \xrightarrow[x \to \infty]{} 0 \tag{4.13}$$

so that

$$\frac{\psi_t + \frac{1}{6} u_x \psi + 4\lambda \psi_x - \frac{1}{3} u \psi_x}{\psi} \xrightarrow[x \to \infty]{} 4\lambda \frac{\psi_x}{\psi} = -4\lambda\kappa = 4\kappa^3 \tag{4.14}$$

Thus the constant is determined and the evolution of $\psi(x,t)$

with respect to t is obtained from Eq. (4.14) to be

$$\psi_t + \frac{1}{6} u_x \psi + 4\lambda\psi_x - \frac{1}{3} u\psi_x - 4\kappa^3\psi = 0 \qquad (4.15)$$

Multiplying this equation by ψ and integrating over x we obtain

$$\frac{d}{dt} \int_{-\infty}^{\infty} dx \left(\frac{1}{2}\psi^2\right) + \frac{1}{6} \int_{-\infty}^{\infty} dx u_x \psi^2 + 4\lambda \int_{-\infty}^{\infty} dx\psi\psi_x$$

$$- \frac{1}{3} \int_{-\infty}^{\infty} dxu\psi\psi_x - 4\kappa^3 \int_{-\infty}^{\infty} dx\psi^2 = 0$$

or, $$\frac{d}{dt} \int_{-\infty}^{\infty} dx \left(\frac{1}{2}\psi^2\right) + 2\lambda \int_{-\infty}^{\infty} dx \frac{\partial\psi^2}{\partial x} - \frac{1}{6} \int_{-\infty}^{\infty} dx \frac{\partial(u\psi^2)}{\partial x}$$

$$+ \frac{1}{3} \int_{-\infty}^{\infty} dxu_x\psi^2 - 4\kappa^3 \int_{-\infty}^{\infty} dx\psi^2 = 0 \quad (4.16)$$

Let us note that each of the total derivative terms vanishes upon integration. Moreover, we recall from Eq. (4.5) that $u_x\psi^2$ has the form of a total derivative and hence vanishes upon integration also. Thus, if we define,

$$c^{-1}(t) = \int_{-\infty}^{\infty} dx\psi^2(x,t) \qquad (4.17)$$

for the bound states, Eq. (4.16) gives the evolution equation

$$\frac{dc^{-1}(t)}{dt} = 8\kappa^3 c^{-1}(t) \qquad (4.18)$$

so that

$$c^{-1}(t) = c^{-1}(0)e^{8\kappa^3 t} \qquad\qquad (4.19)$$

and

$$c(t) = c(0)e^{-8\kappa^3 t} \qquad\qquad (4.20)$$

This, therefore, determines the evolution of the bound state wave function with t.

ii) **Scattering States:** $\lambda = k^2 \geq 0$

In this case we choose the wave functions to have the asymptotic forms

$$\psi(x,t) \xrightarrow[x\to-\infty]{} e^{ikx} + R(k,t)\, e^{-ikx}$$

$$\qquad\qquad (4.21)$$

$$\psi(x,t) \xrightarrow[x\to\infty]{} T(k,t)\, e^{ikx}$$

Here we are assuming a plane wave incident from the left and R(k,t) and T(k,t) represent the coefficients of reflection and transmission respectively.   Unitarity requires them to satisfy

$$|R(k,t)|^2 + |T(k,t)|^2 = 1 \qquad\qquad (4.22)$$

In this case we can calculate the constant of Eq. (4.11) in the following way. We see from Eq. (4.21) that

$$\frac{\psi_t + \frac{1}{6} u_x \psi + 4\lambda\psi_x - \frac{1}{3} u\psi_x}{\psi}$$

$$\xrightarrow[x \to -\infty]{} \frac{R_t e^{-ikx} + 4\lambda ik\ e^{ikx} - 4\lambda ikR\ e^{-ikx}}{e^{ikx} + R\ e^{-ikx}}$$

$$= \frac{(R_t - 4i\lambda kR)e^{-ikx} + 4i\lambda k\ e^{ikx}}{e^{ikx} + R\ e^{-ikx}} \qquad (4.23)$$

For this to be a constant it is clear that

$$R_t - 4i\lambda kR = 4i\lambda kR$$

$$\text{or,} \quad R_t = 8i\lambda kR = 8ik^3 R$$

$$\text{or,} \quad R(k,t) = R(k,o)e^{8ik^3 t} \qquad (4.24)$$

The constant, then, is easily seen from Eq. (4.23) to be

$$\frac{\psi_t + \frac{1}{6} u_x \psi + 4\lambda\psi_x - \frac{1}{3} u\psi_x}{\psi} = 4i\lambda k = 4ik^3 \qquad (4.25)$$

Evaluating the expression of Eq. (4.11) as $x \to \infty$, on the other hand, we obtain

$$\frac{\psi_t + \frac{1}{6} u_x \psi + 4\lambda\psi_x - \frac{1}{3} u\psi_x}{\psi} \xrightarrow[x\to\infty]{} 4ik^3$$

or, $\quad \dfrac{T_t e^{ikx} + 4i\lambda kT\, e^{ikx}}{T\, e^{ikx}} = 4ik^3$

or, $\quad T_t + 4ik^3 T = 4ik^3 T$

or, $\quad T_t = 0$

or, $\quad T(k,t) = T(k,o)$ $\hspace{4cm}$ (4.26)

Thus we have determined that if the potential in the Schrödinger equation of Eq. (4.1) evolves according to the KdV equation, then the coefficients of reflection and transmission evolve in a simple way given by

$$R(k,t) = R(k,o)e^{8ik^3t}$$

(4.27)

$$T(k,t) = T(k,o)$$

Furthermore, each of the bound state normalizations also evolves in a simple way, namely,

$$c_n(t) = \left(\int dx\psi_n^2(x,t)\right)^{-1} = c_n(0)e^{-8\kappa_n^3 t} \hspace{2cm} (4.28)$$

for each bound state with energy $-\kappa_n^2$.

Gel'fand-Levitan Equation:

A given potential in the Schrödinger equation uniquely determines the coefficients of reflection and transmission as well as the bound states which are specified by the c_n's and the κ_n's. Conversely, if we are given these quantities, we can also determine the potential uniquely. Thus philosophically, it is clear that a given initial value for the KdV equation $u(x,o)$ determines $R(k,o)$, $T(k,o)$, κ_n and $c_n(o)$ uniquely. But the evolution of these quantities are already determined from Eqs. (4.9), (4.27) and (4.28) in a simple way. So, we know these quantities at any later time and hence can determine the potential $u(x,t)$ at any later time. This, on the other hand, is the solution of the KdV equation corresponding to the initial value $u(x,o)$.

The actual determination of the potential from a knowledge of the bound states and the coefficients of reflection and transmission is achieved by solving the Gel'fand-Levitan equation. Let $K(x,y)$ be the solution of the equation

$$K(x,y) + B(x+y) + \int_{x}^{\infty} dz K(x,z)B(y+z) = 0 \quad , \quad y \geq x \quad (4.29)$$

with

$$B(x) = \frac{1}{2\pi} \int_{-\infty}^{\infty} dk R(k)e^{ikx} + \sum_{n=1}^{N} c_n e^{-\kappa_n x} \quad (4.30)$$

where N is the total number of bound states of the system. Then the potential is obtained from a knowledge of $K(x,y)$ as

$$\frac{1}{6} u(x) = 2 \frac{\partial}{\partial x} K(x,x) \qquad (4.31)$$

Let us note here that each of the quantities in this integral equation can in principle depend on passive parameters such as t in the case of the KdV equation. Let us also note here that this equation is not in general soluble in closed form. However, if the potential is reflectionless, then closed form solutions exist as we will discuss in an example next.

Example:

To see how this method works, let us take the initial configuration of the KdV solution to be the one soliton configuration. Namely, let (see Eq. (2.24) with c=4)

$$u(x,o) = 12 \text{sech}^2 x \qquad (4.32)$$

so that the Schrödinger equation for t=0 becomes

$$\frac{\partial^2 \psi}{\partial x^2} + \frac{1}{6} u(x,o) \ \psi = -\lambda \psi$$

or, $$\frac{\partial^2 \psi}{\partial x^2} + 2 \text{sech}^2 x \ \psi = -\lambda \psi \qquad (4.33)$$

As we have seen earlier, (see Eqs. (2.64) and (2.65)) this potential supports exactly one bound state with energy

EXAMPLE   87

$$\kappa^2 = -\lambda = 1 \tag{4.34}$$

so that $\kappa = \pm 1$. The bound state solution can be easily determined to be

$$\psi(x) = \frac{1}{2} \text{ sechx} \tag{4.35}$$

We have chosen this particular normalization of the wave function so that asymptotically its behavior is as given in Eq. (4.12), namely,

$$\psi(x) \xrightarrow[x \to -\infty]{} e^x$$

$$\tag{4.36}$$

$$\psi(x) \xrightarrow[x \to \infty]{} e^{-x}$$

That this is the solution of the Schrödinger equation with the correct eigenvalue can be checked as follows.

$$\frac{\partial^2 \psi}{\partial x^2} + 2\text{sech}^2 x \; \psi$$

$$= -\frac{1}{2} \text{ sechx}(1 - 2\tanh^2 x) + 2\text{sech}^2 x \; \frac{1}{2} \text{ sechx}$$

$$= -\frac{1}{2} \text{ sechx}(2\text{sech}^2 x - 1) + \text{sech}^3 x$$

$$= \frac{1}{2} \text{ sechx} = \psi \tag{4.37}$$

The normalization constant in this case can be determined from the definition (4.17) to be

$$c(0) = \left(\int_{-\infty}^{\infty} dx \psi^2(x)\right)^{-1} = \left(\frac{1}{4} \int_{-\infty}^{\infty} dx \ sech^2 x\right)^{-1}$$

$$= \left(\frac{1}{2} \int_{0}^{\infty} dx \ sech^2 x\right)^{-1} = \left(\frac{1}{2} \int_{0}^{1} d(\tanh x)\right)^{-1}$$

$$= 2 \tag{4.38}$$

We know its time evolution from Eq. (4.28) to be

$$c(t) = c(0)e^{-8t} = 2e^{-8t} \tag{4.39}$$

Let us also recall following our discussion of Chapter 2 that the soliton potentials are reflectionless so that

$$R(k,o) = 0 \tag{4.40}$$

From Eq. (4.27) we, then, conclude that

$$R(k,t) = R(k,o)e^{8ik^3 t} = 0 \tag{4.41}$$

That is, the potential remains reflectionless for all values of t. Note that the unitarity relation, Eq. (4.22), in this case, simply becomes

$$|T(k,t)|^2 = |T(k,o)|^2 = 1 \tag{4.42}$$

EXAMPLE   89

Thus the transmission coefficient is a pure phase. However, since this does not enter the Gel'fand-Levitan equation we do not undertake a detailed investigation of its structure. The integral equation of (4.29) in this simple case becomes

$$K(x,y,t) + B(x+y,t) + \int_{x}^{\infty} dz\, K(x,z,t)B(y+z,t) = 0$$

where   (4.43)

$$B(x,t) = c(t)e^{-x} = 2e^{-8t-x}$$

Thus written out explicitly, the equation takes the form

$$K(x,y,t) + 2e^{-8t-x-y} + 2 \int_{x}^{\infty} dz\, K(x,z,t)e^{-8t-y-z} = 0 \quad (4.44)$$

It is obvious that

$$K(x,y,t) = \omega(x,t)e^{-y} \qquad (4.45)$$

Putting this form, into Eq. (4.44), we obtain

$$\omega(x,t) + 2e^{-8t-x} + 2\omega(x,t) \int_{x}^{\infty} dz\, e^{-2z-8t} = 0$$

or, $\quad \omega(x,t)(1+e^{-8t-2x}) = -2e^{-8t-x}$

or, $\quad \omega(x,t) = -\dfrac{2e^{-8t-x-y}}{1+e^{-8t-2x}} \qquad (4.46)$

Therefore,

$$K(x,y,t) = \omega(x,t)e^{-y}$$

$$= -\frac{2e^{-8t-x-y}}{1+e^{-8t-2x}} \tag{4.47}$$

so that

$$K(x,x,t) = -\frac{2e^{-8t-2x}}{1+e^{-8t-2x}} = -2 + \frac{2}{1+e^{-8t-2x}} \tag{4.48}$$

From Eq. (4.31), we then obtain

$$\frac{1}{6} u(x,t) = 2 \frac{\partial}{\partial x} K(x,x,t)$$

$$= \frac{8e^{-8t-2x}}{(1+e^{-8t-2x})^2} = \frac{8}{(e^{4t+x}+e^{-4t-x})^2}$$

$$= 2\text{sech}^2(x+4t)$$

$$\text{or,} \quad u(x,t) = 12\text{sech}^2(x+4t) \tag{4.49}$$

This is, of course, the form of the solution we had obtained earlier (see Eq. (2.24)). It represents only a left moving wave. The significance of this method, however, lies in the fact that it can be applied to any complicated initial configuration.

References:

Gardner, C. S., J. M. Greene, M. D. Kruskal and R. M.
    Miura, Phys. Rev. Lett. $\underline{19}$, 1095 (1967).

Gel'fand, I. M. and B. M. Levitan, Amer. Math. Soc.
    Trans. (2) $\underline{1}$, 253 (1955).

Zakharov, V. E. and L. D. Faddeev, Func. Anal. Appl. $\underline{5}$,
    280 (1971).

CHAPTER 5

INVERSE SCATTERING THEORY

The results of Gardner, Greene, Kruskal and Miura, described in the last chapter, are interesting but they leave many questions unanswered. For example, it is unclear why the dynamics of the nonlinear system should be controlled by a linear system. Secondly, we do not understand whether the simple evolution of the various quantities associated with the linear Schrödinger equation should have a deep origin. Finally, although the inverse scattering method gives a solution of the nonlinear equation, we do not quite see how one can obtain the conserved quantities and hence conclude about the integrability of the system in this formalism. In this chapter, we will try to answer some of the questions postponing a discussion on the origin of the linear Schrödinger system to a latter chapter. Our investigation would be completely in the framework of scattering and inverse scattering theory. Consequently, we will begin by recapitulating briefly the essential details of scattering in one dimension. The rigorous details can be found in many excellent text books as well as review articles.

Scattering in One Dimension:

Let us consider the scattering problem in one dimension for the equation

$$\frac{\partial^2 \psi}{\partial x^2} + (q(x)+k^2)\psi = 0 \qquad (5.1)$$

We assume that the potential q(x) is real and falls off fast enough at infinity. We do not concern ourselves with the explicit rigorous bounds on the fall off behavior of the potential which can be found in the literature. Rather, we assume that the problem is well defined and that a solution to the scattering problem exists. Let us also note here that the Schrödinger system associated with the KdV equation, given in Eq. (4.1), is obtained simply by identifying

$$q(x) = \frac{1}{6} u(x) \qquad\qquad (5.2)$$

Let us consider a plane wave incident from the left. Then we would have an asymptotic wave function of the form

$$\psi(x,k) \xrightarrow[x \to -\infty]{} e^{ikx} + R(k)e^{-ikx}$$

$$\qquad\qquad\qquad\qquad (5.3)$$

$$\xrightarrow[x \to \infty]{} T(k)e^{ikx}$$

As we know $R(k)$ and $T(k)$ are the coefficients of reflection and transmission respectively.

To understand the analytic behavior of $R(k)$ and $T(k)$, let us introduce the Jost functions. These are solutions of Eq. (5.1) which satisfy the boundary conditions

$$f(x,k) \xrightarrow[x \to \infty]{} e^{ikx}$$

(5.4)

$$g(x,k) \xrightarrow[x \to -\infty]{} e^{-ikx}$$

From the form of the Schrödinger equation, Eq. (5.1), we see that for real k, if $f(x,k)$ is a solution, then so is $f^*(x,k)$. From the asymptotic form, on the other hand, we see that

$$f^*(x,k) = f(x,-k)$$

(5.5)

and hence is linearly independent of $f(x,k)$. This is also easily seen by calculating the Wronskian which is independent of x and hence can be evaluated from the asymptotic forms as

$$[f(x,k),f(x,-k)] = \frac{\partial f(x,k)}{\partial x} f(x,-k) - f(x,k) \frac{\partial f(x,-k)}{\partial x}$$

$$= 2ik$$

(5.6)

Similarly, it is also easy to see that

$$g^*(x,k) = g(x,-k)$$

is a solution linearly independent of $g(x,k)$ and that

$$[g(x,k),g(x,-k)] = -2ik$$

(5.7)

Since the Schrödinger equation is a second order differential equation, any solution can be expressed as a linear combination of the pairs $f(x,k)$ and $f(x,-k)$ or $g(x,k)$ and $g(x,-k)$. In particular, we then have

$$f(x,k) = a(k)g(x,-k) + b(k)g(x,k)$$

$$g(x,k) = \tilde{a}(k)f(x,-k) + \tilde{b}(k)f(x,k)$$

(5.8)

The coefficient functions can be easily determined from Eq. (5.8) in terms of the Wronskians as

$$a(k) = \tilde{a}(k) = \frac{1}{2ik} [f(x,k),g(x,k)]$$

$$b(k) = -\frac{1}{2ik} [f(x,k),g(x,-k)]$$

(5.9)

$$\tilde{b}(k) = -\frac{1}{2ik} [f(x,-k),g(x,k)]$$

It follows, therefore, that

$$b(k) = -\tilde{b}(-k)$$

(5.10)

Let us also note that because

$$f^*(x,k) = f(x,-k) \quad \text{and} \quad g^*(x,k) = g(x,-k)$$

the coefficient functions must satisfy

$$a^*(k) = a(-k)$$

$$b^*(k) = b(-k)$$

(5.11)

Finally, the consistency of the relations in Eq. (5.8) further requires that

$$b(k)\widetilde{b}(k) + a(k)\widetilde{a}(-k) = 1$$

$$\text{or,} \quad |a(k)|^2 = 1 + |b(k)|^2 \tag{5.12}$$

It is clear from Eq. (5.12) that $a(k)$ cannot have any zero for any real value of k. It is also clear that if $a(k)$ diverges for any real value of k, then $b(k)$ must also diverge at that point in such a way that

$$\left|\frac{b(k)}{a(k)}\right| = 1 \tag{5.13}$$

We note here that while Eq. (5.12) is reminiscent of the unitarity constraint involving the coefficients of reflection and transmission, given in Eq. (4.22), it is not quite the same.

Analytic Behavior of Scattering Coefficients:

To make contact with $R(k)$ and $T(k)$, let us note that the actual wave function describing the scattering can also be expressed in terms of the Jost functions. In fact, we immediately see from Eq. (5.3) and (5.4) that we can write

$$\psi(x,k) = g(x,-k) + R(k)g(x,k)$$

$$= T(k)f(x,k) \tag{5.14}$$

so that the asymptotic behavior indeed holds. But then substituting the form of f(x,k) from relation Eq. (5.8), we
have

$$T(k) = \frac{1}{a(k)}$$

$$R(k) = \frac{b(k)}{a(k)}$$

(5.15)

It now follows from Eq. (5.12) that

$$|R(k)|^2 + |T(k)|^2 = \left|\frac{b(k)}{a(k)}\right|^2 + \left|\frac{1}{a(k)}\right|^2 = 1 \quad (5.16)$$

This is, of course, the statement about conservation of
probability given in Eq. (4.22).

Let us recognize from our earlier discussion following
Eq. (5.12) that both the coefficients of reflection and
transmission are well defined for all real values of k. To
determine their complete analytic behavior, we extend the
domain of their argument to complex values of k. The Jost
functions are analytic in the upper half of the complex
k-plane where Imk > 0. Consequently, the coefficient
functions, a(k) and b(k), are also analytic in the upper
half of the complex k-plane. Therefore, we see from the
forms of R(k) and T(k) that they are singular only if a(k)
vanishes. But as we have argued before this cannot happen
for any real value of k. Hence let us assume that a(k)
vanishes for some complex value of k in the upper half
plane. If

$$a(k) = 0 \quad \text{for } k = k_o \; , \quad \text{Im } k_o > 0 \tag{5.17}$$

then from the definition of Eq. (5.9), namely,

$$a(k) = \frac{1}{2ik} [f(x,k),g(x,k)]$$

we see that $f(x,k_o)$ and $g(x,k_o)$ must be linearly related. That is, let

$$f(x,k_o) = b_o g(x,k_o) \tag{5.18}$$

From Eq. (5.8), we recognize that we can think of b_o as the analytic continuation of $b(k)$ to $k=k_o$. On the other hand, we note from Eq. (5.4) that since $\text{Im } k_o > 0$, $f(x,k_o)$ and $g(x,k_o)$ both vanish asymptotically, namely,

$$f(x,k_o) \xrightarrow[x\to\infty]{} e^{-(\text{Im } k_o)x}$$

$$g(x,k_o) \xrightarrow[x\to-\infty]{} e^{(\text{Im } k_o)x} \tag{5.19}$$

Thus we see that if $a(k_o) = 0$, then we have a normalizeable eigenfunction of the Schrödinger equation corresponding to the eigenvalue k_o^2. On the other hand, since the Schrödinger operator of Eq. (5.1) is Hermitian, the eigenvalue must be real and since k_o cannot be real, it must be pure imaginary. That is,

$$k_0 = i\kappa_0 \tag{5.20}$$

We recognize this as the position of a bound state. Consequently, we conclude that $a(k)$ can have finitely many zeroes along the positive imaginary axis corresponding to the location of the bound states. One can also show that these are simple zeroes. In fact, from the definition

$$a(k) = \frac{1}{2ik} [f(x,k), g(x,k)]$$

we see that

$$\frac{da(k)}{dk}\bigg|_{k=i\kappa_0} = \Big\{ \frac{1}{2ik} \Big[\frac{\partial f(x,k)}{\partial k} , g(x,k) \Big]$$

$$+ \frac{1}{2ik} [f(x,k) , \frac{\partial g(x,k)}{\partial k}] \Big\} \bigg|_{k=i\kappa_0} \tag{5.21}$$

Here we have used the fact that the Wronskian of $f(x,k)$ and $g(x,k)$ vanishes for $k = i\kappa_0$. Furthermore, let us note that $f(x,k)$ and $g(x,k)$ satisfy the Schrödinger equation

$$\frac{\partial^2 f(x,k)}{\partial x^2} + (q(x)+k^2)f(x,k) = 0 \tag{5.22}$$

$$\frac{\partial^2 g(x,k)}{\partial x^2} + (q(x)+k^2)g(x,k) = 0 \tag{5.23}$$

Differentiating these with respect to k, we have

$$\frac{\partial^2}{\partial x^2}\left(\frac{\partial f(x,k)}{\partial k}\right) + (q(x)+k^2)\left(\frac{\partial f(x,k)}{\partial k}\right) + 2kf(x,k) = 0 \quad (5.24)$$

$$\frac{\partial^2}{\partial x^2}\left(\frac{\partial g(x,k)}{\partial k}\right) + (q(x)+k^2)\left(\frac{\partial g(x,k)}{\partial k}\right) + 2kg(x,k) = 0 \quad (5.25)$$

Multiplying Eq. (5.22) with $\frac{\partial g(x,k)}{\partial k}$ and Eq. (5.25) with $f(x,k)$ and subtracting one from the other we obtain

$$\frac{\partial}{\partial x}\left(\frac{\partial g(x,k)}{\partial k}\frac{\partial f(x,k)}{\partial x} - f(x,k)\frac{\partial}{\partial x}\left(\frac{\partial g(x,k)}{\partial k}\right)\right)$$

$$- 2kf(x,k)g(x,k) = 0$$

or, $\frac{\partial}{\partial x}\left[f(x,k), \frac{\partial g(x,k)}{\partial k}\right] = 2kf(x,k)g(x,k)$ (5.26)

Integrating this with respect to x we obtain

$$\left[f(x,k), \frac{\partial g(x,k)}{\partial k}\right]_{-\infty}^{x} = 2k\int_{-\infty}^{x}dx'f(x',k)g(x',k) \quad (5.27)$$

Similarly, from equations (5.23) and (5.24), we obtain

$$\left[\frac{\partial f(x,k)}{\partial k}, g(x,k)\right]_{x}^{\infty} = -2k\int_{x}^{\infty}dx'f(x',k)g(x',k) \quad (5.28)$$

Let us note further from Eq. (5.19) that for $k = i\kappa_o$, $f(x,k)$ and its derivatives vanish exponentially as $x \to \infty$ just as $g(x,k)$ and its derivative do for $x \to -\infty$. Thus we obtain

$$\left\{\left[f(x,k), \frac{\partial g(x,k)}{\partial k}\right] + \left[\frac{\partial f(x,k)}{\partial k}, g(x,k)\right]\right\}\bigg|_{k=i\kappa_o}$$

$$= 2i\kappa_o \int_{-\infty}^{\infty} dx f(x,i\kappa_o)g(x,i\kappa_o) \qquad (5.29)$$

Thus we see from Eq. (5.21) that

$$\frac{da(k)}{dk}\bigg|_{k=i\kappa_o} = -\frac{1}{2\kappa_o}(2i\kappa_o)\int_{-\infty}^{\infty} dx f(x,i\kappa_o)g(x,i\kappa_o)$$

$$= -i \int_{-\infty}^{\infty} dx f(x,i\kappa_o)g(x,i\kappa_o)$$

$$\text{or, } i\frac{da(k)}{dk}\bigg|_{k=i\kappa_o} = b_o^{-1}\int_{-\infty}^{\infty} dx \, f^2(x,i\kappa_o) = b_o^{-1}c_o^{-1} \quad (5.30)$$

where c_o is the normalization of the bound state as discussed in Eq. (4.17). This shows that the zero at $k = i\kappa_o$
is a simple zero since neither b_o or c_o vanish for a nontrivial solution. Furthermore, if the spectrum does not
change with the evolution of the potential, then the left
hand side is a constant and consequently, b_o would be
proportional to c_o^{-1}.

We, therefore, conclude from Eqs. (5.11), (5.15), (5.20)
and (5.30) that the coefficients of transmission and
reflection are continuous functions of k, satisfying

$$R^*(k) = R(-k) \qquad T^*(k) = T(-k) \qquad (5.31)$$

Furthermore, they are analytic in the complex upper half plane except at the locations of the bound states, namely $k_n = i\kappa_n$, $n = 1,2,\ldots N$. The singularities of $T(k)$ at the bound states correspond to simple poles with the residues

$$\text{Res } T(k)\Big|_{k=i\kappa_n} = \left(-i \int_{-\infty}^{\infty} dx f(x,i\kappa_n) g(x,i\kappa_n)\right)^{-1}$$

$$= i \, b_n \, c_n \qquad\qquad (5.32)$$

where b_n and c_n's are the generalizations of b_0 and c_0 to other bound states.

It is clear from these properties that we can determine the coefficient of transmission once we know the coefficient of reflection and the bound states. For example, if we know $R(k)$,

$$|T(k)| = (1-|R(k)|^2)^{1/2} \qquad\qquad (5.33)$$

Furthermore, since $T(k)$ is meromorphic we can write

$$T(k) = \exp\left\{\frac{1}{2\pi i} \int_{-\infty}^{\infty} dk' \, \frac{\ell n(1-|R(k')|^2)}{k'-k}\right\} \prod_{n=1}^{N} \frac{k+i\kappa_n}{k-i\kappa_n} \qquad (5.34)$$

The second factor, which arises from the bound state contributions, corresponds to the phase of $T(k)$. Thus we see that the scattering matrix is completely determined once we know $R(k)$, κ_n and c_n. That is why this set is often called the scattering data. Furthermore, just as a potential determines the scattering data and hence the S-matrix uniquely, the knowledge of the scattering data also can lead to the construction of the potential uniquely. Finally, to see the usefulness of the formula in Eq. (5.34), let us note that if the potential is reflectionless, then

$$R(k) = 0$$

and hence

$$T(k) = \prod_{n=1}^{N} \frac{k+i\kappa_n}{k-i\kappa_n} \tag{5.35}$$

is a pure phase.

To fix our ideas, let us consider the one soliton example, studied in Chapter 4, in some detail. In this case, we obtained that the potential is reflectionless and supports only one bound state at

$$k = i$$

so that from Eq. (5.35), we obtain

$$T(k) = \frac{1}{a(k)} = \frac{k+i}{k-i} \tag{5.36}$$

It follows, then, that

$$i \left. \frac{da(k)}{dk} \right|_{k=i} = \frac{1}{2} \tag{5.37}$$

We also obtain from Chapter 4 (see Eq. (4.35)), that the bound state wave function has the form

$$\psi(x,t) = \frac{1}{2} \operatorname{sech}(x+4t)$$

which satisfies

$$\psi(x,t) \xrightarrow[x \to \pm\infty]{} e^{\mp(x+4t)}$$

and   (5.38)

$$\int_{-\infty}^{\infty} dx \, \psi^2(x,t) = \frac{1}{2}$$

The Jost functions can be written in terms of $\psi(x,t)$ as

$$f(x,k=i) = e^{4t}\psi(x,t) \xrightarrow[x \to \infty]{} e^{-x}$$

(5.39)

$$g(x,k=i) = e^{-4t}\psi(x,t) \xrightarrow[x \to -\infty]{} e^{x}$$

Note that

$$f(x,k=i) = e^{8t}g(x,k=i)$$

so that   (5.40)

$$b(k=i) = e^{8t}$$

We recognize this to be $2c^{-1}(t)$ from Eq. (4.39). Further-more,

$$\int_{-\infty}^{\infty} dx\; f^2(x,k=i) = e^{8t} \int_{-\infty}^{\infty} dx\; \psi^2(x,t) = \frac{1}{2}\, e^{8t} = c^{-1}(t) \quad (5.41)$$

From Eqs. (5.37), (5.40) and (5.41) we can verify in a straightforward manner that the relation (5.30) holds.

Action-Angle Variables for KdV:

As we have noted in Eq. (5.2), we can specialize this discussion to KdV by choosing

$$q(x) = \frac{1}{6}\, u(x)$$

so that the Schrödinger equation becomes

$$\frac{\partial^2 \psi}{\partial x^2} + (\frac{1}{6}\, u(x)+k^2)\psi = 0$$

Instead of dealing with the scattering coefficients directly, let us analyze $a(k)$ and $b(k)$. It is clear that these coefficients are functionals of the potential. Therefore, we could calculate their variation with respect to the potential in the following way. Since the Jost functions satisfy the Schrödinger equation, we have

$$\frac{\partial^2 f(x,k)}{\partial x^2} + (\frac{1}{6} u(x)+k^2)f(x,k) = 0$$

$$(5.42)$$

$$\frac{\partial^2 g(x,k)}{\partial x^2} + (\frac{1}{6} u(x)+k^2)g(x,k) = 0$$

Let us assume infinitesimal variations

$$u(x) \rightarrow u(x) + \delta u(x)$$

$$f(x,k) \rightarrow f(x,k) + \delta f(x,k)$$

$$g(x,k) \rightarrow g(x,k) + \delta g(x,k)$$

subject to the conditions $$(5.43)$$

$$\delta u \xrightarrow[|x| \rightarrow \infty]{} 0$$

$$\delta f(x,k) \xrightarrow[x \rightarrow \infty]{} 0$$

$$\delta g(x,k) \xrightarrow[x \rightarrow -\infty]{} 0$$

Keeping linear terms in the variation of Eq. (5.42), we
obtain

$$\frac{\partial^2 \delta f}{\partial x^2} + (\frac{1}{6} u(x)+k^2)\delta f + \frac{1}{6} \delta u f = 0$$

$$(5.44)$$

$$\frac{\partial^2 \delta g}{\partial x^2} + (\frac{1}{6} u(x)+k^2)\delta g + \frac{1}{6} \delta u g = 0$$

Multiplying the first of Eq. (5.42) with δg and the last of Eq. (5.44) with f and subtracting one from the other we have

$$\frac{\partial}{\partial x} \left(\frac{\partial f}{\partial x} \delta g - f \frac{\partial \delta g}{\partial x} \right) - \frac{1}{6} \delta u f g = 0$$

or, $[f, \delta g]_{-\infty}^{x} = \frac{1}{6} \int_{-\infty}^{x} dx' \delta u(x') f(x', k) g(x', k)$

or, $\left[f(x,k) , \frac{\delta g(x,k)}{\delta u(x)} \right] = \frac{1}{12} f(x,k) g(x,k)$ (5.45)

In deriving Eq. (5.45) we have used the asymptotic conditions in Eq. (5.43) and the formula

$$\int_{0}^{\infty} dx \delta(x) f(x) = \frac{1}{2} f(0) = \int_{-\infty}^{0} dx \, \delta(x) f(x)$$

Similarly, from the other pair of equations in (5.42) and (5.44) we obtain

$$\left[\frac{\delta f(x,k)}{\delta u(x)} , g(x,k) \right] = \frac{1}{12} f(x,k) g(x,k)$$ (5.46)

Then, from

$$a(k) = \frac{1}{2ik} [f(x,k), g(x,k)]$$

and using Eqs. (5.45) and (5.46), we have

$$\frac{\delta a(k)}{\delta u(x)} = \frac{1}{2ik} \left[\frac{\delta f(x,k)}{\delta u(x)} , g(x,k) \right]$$

$$+ \frac{1}{2ik} \left[f(x,k) , \frac{\delta g(x,k)}{\delta u(x)} \right]$$

$$\text{or,} \quad \frac{\delta a(k)}{\delta u(x)} = \frac{1}{12ik} f(x,k)g(x,k) \qquad (5.47)$$

Similarly, we can also calculate from

$$b(k) = - \frac{1}{2ik} [f(x,k),g(x,-k)]$$

that

$$\frac{\delta b(k)}{\delta u(x)} = - \frac{1}{12ik} f(x,k)g(x,-k) \qquad (5.48)$$

We could, therefore, calculate the Poisson brackets between various functionals of a(k) and b(k) from the relation, Eq. (1.53)

$$\{F,G\} = \frac{1}{2} \int_{-\infty}^{\infty} dx \left(\frac{\delta F}{\delta u(x)} \frac{\partial}{\partial x} \frac{\delta G}{\delta u(x)} - \frac{\partial}{\partial x} \frac{\delta F}{\delta u(x)} \frac{\delta G}{\delta u(x)} \right)$$

The explicit calculations of the Poisson brackets are given in the appendix and we merely state here that the set

$$P(k) = - \frac{144k}{\pi} \log|a(k)|$$

$$(5.49)$$

$$Q(k) = \arg b(k)$$

constitute a canonical set of variables satisfying for
$k, k' > 0$

$$\{P(k), P(k')\} = 0 = \{Q(k), Q(k')\}$$

$$\{Q(k), P(k')\} = \delta(k-k') \tag{5.50}$$

In addition to the scattering states, we also have the
bound states characterized by κ_n and c_n. One can calculate
the Poisson brackets between these variables in a similar
manner and it can be shown that

$$p_n = 144 \, \kappa_n^2$$

$$q_n = \frac{1}{2} \log |b_n| \tag{5.51}$$

where

$$b_n = -ic_n^{-1} \left/ \frac{da(k)}{dk} \right|_{k=i\kappa_n} \qquad n = 1, 2, \ldots .N$$

also constitute a canonical set under the KdV Poisson
bracket structure. Together the set $(P(k), p_n, Q(k), q_n)$
constitute the canonical coordinates of KdV. It follows
from this that

$$\{\log a(k), \log a(k')\} = 0 \tag{5.52}$$

Consequently, $\log a(k)$ must somehow contain the conserved
quantities.

In fact, let us note from Eq. (5.34) that

$$\log a(k) = -\frac{1}{2\pi i} \int_{-\infty}^{\infty} dk' \frac{\log(1-|R(k')|^2)}{k'-k}$$

$$+ \sum_{m=1}^{N} \log \frac{k-i\kappa_m}{k+i\kappa_m} \tag{5.53}$$

and since $|R(k)|$ is small for large k, this has the asymptotic expansion

$$\log a(k) \xrightarrow[k\to\infty]{} \sum_{n=0}^{\infty} \frac{C_{2n+1}}{k^{2n+1}} \tag{5.54}$$

where

$$C_{2n+1} = \frac{1}{2\pi i} \int_{-\infty}^{\infty} dk \, k^{2n} \log(1-|R(k)|^2) - \frac{2}{2n+1} \sum_{m=1}^{N} (i\kappa_m)^{2n+1}$$

$$= \frac{1}{2\pi i} \int_{-\infty}^{\infty} dk \, k^{2n} \log|T(k)|^2 - \frac{2}{2n+1} \sum_{m=1}^{N} (i\kappa_m)^{2n+1}$$

$$= -\frac{i}{72} \int_{0}^{\infty} dk \, k^{2n-1} P(k) - \frac{2}{2n+1} (\frac{i}{12})^{2n+1} \sum_{m=1}^{N} (P_m)^{\frac{2n+1}{2}} \tag{5.55}$$

It is clear that the C_{2n+1}'s involve only the momentum variables and hence must be in involution. To make contact with the more familiar form of the conserved quantities we proceed as follows.

We know from Eq. (5.4) that

$$f(x,k) \xrightarrow[x\to\infty]{} e^{ikx}$$

Hence

$$\log f(x,k) \xrightarrow[x\to\infty]{} ikx$$

$$\text{or,} \quad \chi(x,k) = \log f(x,k) - ikx \xrightarrow[x\to\infty]{} 0 \qquad (5.56)$$

Similarly, from the definition in Eq. (5.9)

$$a(k) = \frac{1}{2ik}\left(\frac{\partial f}{\partial x} g - f \frac{\partial g}{\partial x}\right)$$

$$\xrightarrow[x\to-\infty]{} \frac{1}{2ik}\left(\frac{\partial f}{\partial x} e^{-ikx} + ikfe^{-ikx}\right)$$

$$= \frac{1}{2ik} f(x,k)e^{-ikx}\left(\frac{1}{f}\frac{\partial f}{\partial x} + ik\right)$$

$$= \frac{1}{2ik} f(x,k)e^{-ikx}\left(\frac{\partial}{\partial x}(\ell nf - ikx) + 2ik\right)$$

$$= f(x,k)e^{-ikx}\left(1 + \frac{1}{2ik}\frac{\partial\chi(x,k)}{\partial x}\right) \qquad (5.57)$$

Taking logarithm of both sides we conclude that

$$\chi(x,k) \xrightarrow[x\to-\infty]{} \log a(k)$$

Consequently, defining

$$\xi(x,k) = \frac{\partial\chi(x,k)}{\partial x} = \frac{1}{f}\frac{\partial f}{\partial x} - ik \qquad (5.58)$$

we see that

$$\log a(k) = - \int_{-\infty}^{\infty} dx\, \xi(x,k) \tag{5.59}$$

On the other hand, since $f(x,k)$ satisfies (see Eq. (5.38))

$$\frac{\partial^2 f(x,k)}{\partial x^2} + (\frac{1}{6}\, u(x)+k^2)f(x,k) = 0$$

we obtain

$$\frac{\partial \xi}{\partial x} = \frac{\partial}{\partial x} \left(\frac{1}{f}\, \frac{\partial f}{\partial x} - ik\right)$$

$$= -\xi^2 - 2ik\xi - \frac{1}{6}\, u$$

or, $\quad \xi^2 + \frac{\partial \xi}{\partial x} + 2ik\xi + \frac{1}{6}\, u = 0 \tag{5.60}$

That is, ξ satisfies a generalized Riccati equation. Expanding ξ as a power series in inverse powers of k we have

$$\xi(x,k) = \sum_{n=0}^{\infty} \frac{\xi_n(x)}{(2ik)^n} \tag{5.61}$$

where the ξ_n's must satisfy the recursion relation

$$\xi_n + \frac{d\xi_{n-1}}{dx} + \sum_{m=1}^{n-1} \xi_{n-m-1}\xi_m = 0 \tag{5.62}$$

with

$$\xi_0 = 0$$

$$\xi_1 = -\frac{1}{6} u \tag{5.63}$$

Thus, recursively, we see that

$$\xi_2 = -\frac{d\xi_1}{dx} = \frac{1}{6} u_x$$

$$\xi_3 = -\left(\frac{d\xi_2}{dx} + \xi_1^2\right) = -\frac{1}{6} u_{xx} - \frac{1}{36} u^2 \tag{5.64}$$

and so on. Up to normalization constants we see that these are the conserved densities we had calculated in Eqs. (3.36) -(3.41). In fact, we can show that

$$\xi_n = \frac{(-i)^{n+1}}{6} v_{n-1} \tag{5.65}$$

so that following Eq. (3.44) we can write

$$H_n = 3(-1)^n \int_{-\infty}^{\infty} dx \, v_{2n} = -18 \int_{-\infty}^{\infty} dx \, \xi_{2n+1} \tag{5.66}$$

We see from Eqs. (5.59) and (5.61) that we can write, asymptotically,

$$\log a(k) = - \sum_{n=0}^{\infty} \frac{1}{(2ik)^{2n+1}} \int_{-\infty}^{\infty} dx \; \xi_{2n+1}(x) \qquad (5.67)$$

But we also know from Eq. (5.54) that for large k

$$\log a(k) = \sum_{n=0}^{\infty} \frac{1}{k^{2n+1}} \; c_{2n+1}$$

Thus, comparing, we obtain

$$c_{2n+1} = - \frac{1}{(2i)^{2n+1}} \int_{-\infty}^{\infty} dx \xi_{2n+1}(x) \qquad (5.68)$$

where the familiar conserved densities are calculated recursively. The conserved quantities of Eq. (5.66) are then given in terms of the action variables as

$$H_n = -18 \int_{-\infty}^{\infty} dx \xi_{2n+1} = 18(2i)^{2n+1} c_{2n+1}$$

$$= (-1)^n \, 2^{2n-1} \int_{0}^{\infty} dk \; k^{2n-1} P(k)$$

$$+ \frac{1}{2n+1} \, (\tfrac{1}{6})^{2n-1} \sum_{m=1}^{N} (p_m)^{\frac{2n+1}{2}} \qquad (5.69)$$

In particular, the KdV hamiltonian takes the form

$$H_{KdV} = H_2$$

$$= 8 \int_{0}^{\infty} dk \; k^3 P(k) + \frac{1}{5} \, (\tfrac{1}{6})^3 \sum_{m=1}^{N} (p_m)^{5/2} \qquad (5.70)$$

The time evolution of the canonical variables can now be simply calculated using the Poisson brackets of Eqs. (5.50) and (5.51).

$$\frac{dP(k,t)}{dt} = \{P(k,t),H_{KdV}\} = 0$$

$$\frac{dp_n(t)}{dt} = \{p_n(t),H_{KdV}\} = 0$$

$$\frac{dQ(k,t)}{dt} \{Q(k,t),H_{KdV}\} = 8k^3 \qquad (5.71)$$

$$\frac{dq_n(t)}{dt} = \{q_n(t),H_{KdV}\} = 4\kappa_n^3$$

From the definitions of these variables, it then follows that Eq. (5.71) is equivalent to

$$a(k,t) = a(k,o)$$

$$b(k,t) = b(k,o)e^{8ik^3t} \qquad (5.72)$$

$$b_n(t) = b_n(o)e^{8\kappa_n^3t}$$

This is the same evolution as was obtained by Gardner, Greene, Kruskal and Miura (see Eqs. (4.27) and (4.28)). We now understand this simple evolution to be a consequence of the fact that the transformation to the scattering data is the transformation to the action angle variables.

References:

Agranovich, Z. S. and V. A. Marchenko, The Inverse
Problem of Scattering, Gordon and Breach, 1963.

Chadan, K. and P. C. Sabatier, Inverse Problems in
Scattering Theory, Springer-Verlag, 1977.

Faddeev, L. D., J. Math. Phys. $\underline{4}$, 72 (1963).

Faddeev, L. D., J. Sov. Math. $\underline{5}$, 334 (1976).

Kay, I. and H. E. Moses, Nuovo Cim. $\underline{2}$, 276 (1956).

Kay, I. and H. E. Moses, J. Appl. Phys. $\underline{27}$, 1503 (1956).

Novikov, S., S. V. Manakov, L. P. Pitaevskii and V. E.
Zakharov, Theory of Solitons, Consultants Bureau,
1984.

Zakharov, V. E. and L. D. Faddeev, Func. Anal. Appl. $\underline{5}$,
280 (1971).

Zakharov, V. E. and S. V. Manakov, Theo. Math. Phys. $\underline{19}$,
551 (1974).

CHAPTER 6

THE LAX METHOD

So far we have seen that if we can associate an appropriate linear equation with a given nonlinear evolution equation, then the method of inverse scattering applied to the linear system yeidls the solution of the nonlinear equation. The linear system whose eigenvalues do not evolve under the nonlinear flow, therefore, plays a fundamental role. However, we have not as yet understood how the linear equation arises and how to find an appropriate linear equation for a given nonlinear evolution equation.

Origin of the Schrödinger Equation:

There are various ways to understand the origin of the Schrödinger equation of Chapter 4 for the KdV equation. Let us first understand the most intuitive approach before going into a full formal discussion of the Lax theory. We recall that the KdV and the MKdV equations are related through a Riccati relation. Namely, if (see Eqs. (3.17) and (3.18))

$$u(x,t) = v^2(x,t) + i\sqrt{6}\, \frac{\partial v(x,t)}{\partial x} \qquad (6.1)$$

then

$$\frac{\partial u}{\partial t} - u\, \frac{\partial u}{\partial x} - \frac{\partial^3 u}{\partial x^3} = 0 \qquad (6.2)$$

implies

$$\frac{\partial v}{\partial t} - v^2 \frac{\partial v}{\partial x} - \frac{\partial^3 v}{\partial x^3} = 0 \qquad\qquad (6.3)$$

In fact, we can take advantage of the Galilean invariance of the KdV equation, Eq. (1.42), and define a generalized Riccati relation of the form

$$u(x,t) + 6\lambda = v^2(x,t) + i\sqrt{6} \frac{\partial v(x,t)}{\partial x} \qquad (6.4)$$

Then

$$\frac{\partial u}{\partial t} - u \frac{\partial u}{\partial x} - \frac{\partial^3 u}{\partial x^3} = 0 \qquad\qquad (6.5)$$

would imply

$$\frac{\partial v}{\partial t} - (v^2 - 6\lambda) \frac{\partial v}{\partial x} - \frac{\partial^3 v}{\partial x^3} = 0 \qquad (6.6)$$

As we have discussed earlier, a solution of the generalized MKdV equation gives a solution of the KdV equation through the Riccati relation. However, the converse is not true in general since the Riccati relation is not invertible in general. However, since both the KdV and the MKdV equations are integrable and share the same conserved quantities, we feel intuitively that the solution of one should imply the other and, therefore, we can ask when it is that one should be able to invert the Riccati relation. The simplest way to invert the Riccati relation is to linearize and so we define

$$v(x,t) = i\sqrt{6} \frac{\psi_x}{\psi} \tag{6.7}$$

so that the Riccati relation of Eq. (6.4) becomes

$$(u(x,t)+6\lambda) = - 6 \frac{\psi_{xx}}{\psi}$$

or, $\frac{\partial^2\psi}{\partial x^2} + (\frac{1}{6} u(x,t)+\lambda)\psi = 0$ $\tag{6.8}$

Therefore, if we know ψ which satisfies the above equation then we can invert the Riccati relation. This is, of course, the time independent Schrödinger equation we studied in Eq. (4.1) and must hold for all values of the parameter t. Furthermore, note that since λ, the eigenvalue of the Schrödinger equation or the spectral parameter, enters our discussion through a Galilean transformation, it is independent of the parameter t. Consequently, this explains why the spectral parameter does not evolve with t.

The Schrödinger wave function ψ can be expressed in terms of the MKdV variable v(x,t) as

$$\psi(x,t) = \exp\left(- \frac{i}{\sqrt{6}} \int^x dx'v(x',t)\right) \tag{6.9}$$

so that we can obtain its time evolution to be

$$\psi_t = \psi\left(-\frac{i}{\sqrt{6}} \int^x dx' \frac{\partial v(x',t)}{\partial t}\right) \qquad (6.10)$$

On the other hand, we also know the time evolution of v from Eq. (6.6), namely,

$$\frac{\partial v}{\partial t} = (v^2 - 6\lambda) \frac{\partial v}{\partial x} + \frac{\partial^3 v}{\partial x^3} \qquad (6.11)$$

or, $\frac{\partial v}{\partial t} = \frac{\partial}{\partial x} \left(\frac{1}{3} v^3 - 6\lambda v + \frac{\partial^2 v}{\partial x^2}\right)$

Substituting this into Eq. (6.10), the time evolution of ψ becomes

$$\psi_t = -\frac{i}{\sqrt{6}} \psi\left(\frac{1}{3} v^3(x,t) - 6\lambda v(x,t)\right.$$

$$\left. + \frac{\partial^2 v(x,t)}{\partial x} + const\right) \qquad (6.12)$$

Putting in the form of v in terms of ψ from Eq. (6.7) and using the fact that ψ satisfies the Schrödinger equation, Eq. (6.8), we determine the time evolution of ψ to be

$$\psi_t = \psi_{xxx} + \frac{1}{2} u\psi_x - 3\lambda\psi_x + const. \ \psi$$

That is,

$$\psi_t - \psi_{xxx} - \frac{1}{2} u\psi_x + 3\lambda\psi_x = const. \ \psi$$

or, $\psi_t + \frac{1}{6} u_x \psi + 4\lambda\psi_x - \frac{1}{3} u\psi_x = \text{const. } \psi$ (6.13)

This is indeed the form of the evolution equation we have already used in Eq. (4.11) in the investigation of the inverse scattering problem for the KdV equation.

The Lax Pair:

Let us next try to understand the formal theory due to Lax. The basic question we are interested in is the following. Given a nonlinear evolution equation, we would like to find a linear operator whose eigenvalues are constant under the nonlinear evolution. This is, of course, at the heart of the success of the inverse scattering method. To understand this question better, let us analyze the corresponding question for the linear case. Namely, given a linear evolution equation described by a time independent Hamiltonian H, we would like to construct operators whose expectation values do not change with time. It is clear that if A is an operator with this property, then in the Heisenberg picture, A(t) must be unitarily equivalent to A(0). Namely,

$$U^+(t)A(t)U(t) = A(0)$$ (6.14)

Here U(t) is the time evolution operator with the form (in the present case)

$$U(t) = \exp(-iHt)$$ (6.15)

Differentiating both sides of Eq. (6.14) with respect to t we obtain

$$U^+(t) \left(\frac{\partial A(t)}{\partial t} - i[A(t),H] \right) U(t) = 0$$

$$\text{or,} \quad \frac{\partial A(t)}{\partial t} = i[A(t),H] \qquad\qquad (6.16)$$

That is, for the expectation value of A(t) to be time independent, Eq. (6.16) must be satisfied. Furthermore, we note that we can write

$$\frac{\partial U(t)}{\partial t} = -iHU(t) = BU(t)$$

where   (6.17)

$$B = -iH$$

is an anti Hermitian operator.

Let us next mimic this argument in the case of the nonlinear evolution equation. Let

$$L(u(x,t)) = L(t) \qquad\qquad (6.18)$$

denote the linear operator we are interested in finding. We assume this to be Hermitian and that its eigenvalues are independent of t. For this to be true, there must exist a unitary operator U(t) such that

$$U^+(t)L(t)U(t) = L(0) \tag{6.19}$$

Differentiating both sides with respect to t, we obtain

$$\frac{\partial U^+(t)}{\partial t} L(t)U(t) + U^+(t) \frac{\partial L(t)}{\partial t} U(t)$$

$$+ U^+(t)L(t) \frac{\partial U(t)}{\partial t} = 0 \tag{6.20}$$

Unlike the linear case, here we do not know the form of U(t), but it is clear that because U(t) is unitary

$$U^+(t)U(t) = 1$$

$$\frac{\partial U^+(t)}{\partial t} U(t) + U^+(t) \frac{\partial U(t)}{\partial t} = 0 \tag{6.21}$$

so that as before in Eq. (6.17) we can write

$$\frac{\partial U(t)}{\partial t} = B(t)U(t) \tag{6.22}$$

where B(t) is an operator which must be anti Hermitian. Substituting this back into Eq. (6.20) we obtain

$$U^+(t)\left(\frac{\partial L(t)}{\partial t} - [B(t),L(t)]\right)U(t) = 0$$

or, $\quad \frac{\partial L(t)}{\partial t} = [B(t),L(t)] \tag{6.23}$

Thus we see that for L(t) to be isospectral, it must

satisfy a relation similar to the linear case obtained in
Eq. (6.16). The only difference is that, in the present
case, we do not know the form of B(t). But let us note that
L(t) is linear in u(x,t). Consequently, the left hand side
of Eq. (6.23) would be a multiplicative operator which would
be proportional to the time evolution of u(x,t). It is
clear, therefore, that if we can find a linear operator L(t)
and a second operator B(t), which is not necessarily linear,
such that the commutator [B(t),L(t)] is a multiplicative
operator and is proportional to the evolution of u(x,t)
according to the nonlinear equation, then the eigenvalues of
L(t) would be independent of t. Namely, in such a case the
eigenvalues, λ, of the equation

$$L(t)\psi(t) = -\lambda\psi(t) \qquad\qquad (6.24)$$

would be independent of t. Furthermore, $\psi(t)$ must be
unitarily related to its value at t=0. That is to say,

$$\psi(t) = U(t)\psi(0) \qquad\qquad (6.25)$$

Consequently, its evolution with t would take the form

$$\frac{\partial\psi(t)}{\partial t} = \frac{\partial U(t)}{\partial t}\psi(0) = B(t)\psi(t) \qquad\qquad (6.26)$$

The operators L(t) and B(t), when exist, are known as the
Lax pair corresponding to a given nonlinear evolution
equation and play a fundamental role in determining the
solution.

Specialization to KdV:

Let us next specialize to the case of the KdV equation. In this case, we have seen that

$$L(t) = \frac{\dot{\partial}^2}{\partial x^2} + \frac{1}{6} u(x,t) = D^2 + \frac{1}{6} u(x,t) \qquad (6.27)$$

so that

$$\frac{\partial L(t)}{\partial t} = \frac{1}{6} \frac{\partial u}{\partial t} \qquad (6.28)$$

This is indeed a multiplicative operator and describes the time evolution of u. One can look for a B(t) in a systematic way. Let us note that B(t) must be anti Hermitian and hence must be odd in the number of derivatives. Let us choose the simplest form, namely,

$$B(t) = a \, D \qquad (6.29)$$

where a is a constant. In this case,

$$[B(t),L(t)] = \frac{a(Du(x,t))}{6} = \frac{a}{6} \frac{\partial u}{\partial x} \qquad (6.30)$$

Thus while the commutator is a multiplicative operator, it does not describe the evolution of u(x,t) under the KdV flow. In fact, we see that, with a=1, the Lax equation in the present case describes the chiral particle or wave. Namely,

$$\frac{\partial L}{\partial t} = [B, L]$$

or, $\frac{\partial u}{\partial t} = \frac{\partial u}{\partial x}$ (6.31)

The next obvious choice of B(t) satisfying all the symmetry properties and a homogeneous scaling behavior is

$$B(t) = a_3 D^3 + a_1(Du + uD)$$ (6.32)

where a_1 and a_3 are constants so that

$$[B(t), L(t)] = \left(\frac{a_3}{6} - a_1\right)(D^3 u) + \frac{a_1}{3} u(Du)$$

$$+ \left(\frac{a_3}{2} - 4a_1\right)\left((D^2 u)D + (Du)D^2\right)$$ (6.33)

Thus we see that this commutator is a multiplicative operator if the constants a_1 and a_3 satisfy

$$\frac{a_3}{2} - 4a_1 = 0$$

or, $a_3 = 8a_1$ (6.34)

In this case

$$[B(t), L(t)] = \frac{a_1}{3} (u(Du) + (D^3 u))$$ (6.35)

and we see that with the choice $a_1 = 1/2$, the Lax equation of Eq. (6.23) indeed describes the KdV equation. Namely,

$$\frac{\partial L}{\partial t} = [B,L]$$

$$\text{or,} \quad \frac{\partial u}{\partial t} = u \frac{\partial u}{\partial x} + \frac{\partial^3 u}{\partial x^3} \qquad (6.36)$$

The Lax pair in this case are given by

$$L(t) = D^2 + \frac{1}{6} u$$

$$B(t) = 4D^3 + \frac{1}{2} (Du+uD) \qquad (6.37)$$

Let us note here that the operator $B(t)$ is defined only up to an additive constant since a constant commutes with everything. Allowing for this then, the evolution of ψ under t follows from Eqs. (6.26) and (6.37) to be

$$\frac{\partial \psi}{\partial t} = B\psi$$

$$= \left(4D^3 + \frac{1}{2} (Du+uD) + \text{const}\right)\psi$$

$$= 4\psi_{xxx} + \frac{1}{2} u_x\psi + u\psi_x + \text{const.} \ \psi$$

$$= -\frac{1}{6} u_x\psi + \frac{1}{3} u\psi_x - 4\lambda\psi_x + \text{const.} \ \psi$$

$$\text{or,} \quad \psi_t + \frac{1}{6} u_x\psi + 4\lambda\psi_x - \frac{1}{3} u\psi_x = \text{const.} \ \psi$$

This is precisely the same equation as in Eqs. (4.11) and (6.13).

In fact, we can generalize this construction and choose B(t) to be higher order in the derivatives. Since B(t) must be anti Hermitian, the most general form with a homogeneous scaling behavior can be chosen to be

$$B_m(t) = a\left[D^{2m+1} + \sum_{j=1}^{m}\left(b_j(u)D^{2j-1} + D^{2j-1}b_j(u)\right)\right] \quad (6.38)$$

Here a is an overall constant and $b_j(u)$ represent m arbitrary functionals of u with specific scaling behavior. We also know that $[B_m(t),L(t)]$ is Hermitian and consequently, must have the form

$$[B_m(t),L(t)] = \tilde{a}\,K_m(u) + \sum_{j=1}^{m} D^j c_j(u)D^j \quad (6.39)$$

where $K_m(u)$ is a multiplicative operator. Requiring this commutator to be a multiplicative operator imposes m conditions which determine all the $b_j(u)$ uniquely. Choosing the overall constant appropriately, the Lax equation in this case can be shown to give the mth order KdV equation of Eq. (3.64), namely,

$$\frac{\partial L}{\partial t} = [B_m(t),L(t)]$$

$$\text{or,} \quad \frac{\partial u}{\partial t} = \frac{\partial}{\partial x}\frac{\delta H_{m+1}}{\delta u(x)} \quad (6.40)$$

This shows that the solutions for all the equations in the KdV hierarchy can be obtained from the scattering data of the same Schrödinger equation.

Alternate Construction:

There exists an alternate method of deriving the operators $B_m(t)$, which uses formal operator techniques and is quite elegant. Let us recall that

$$L(t) = D^2 + \frac{1}{6} u \qquad (6.41)$$

Therefore, we can formally define the square root of this operator to be an infinite series in inverse powers of D. That is,

$$L^{\frac{1}{2}}(t) = D + a_0(u) + \sum_{n=1}^{\infty} a_n(u)D^{-n} \qquad (6.42)$$

where each of the $a_n(u)$'s is a functional of u. Their functional forms can be determined up to any order by squaring the formal series and requiring it to be equal L(t) up to that particular order.

Given the form of $L^{\frac{1}{2}}(t)$, we can easily calculate $(L(t))^{\frac{2m+1}{2}}$ simply from the relation

$$(L(t))^{\frac{2m+1}{2}} = L^m(t)L^{\frac{1}{2}}(t) = (D^2 + \frac{1}{6}u)^m L^{\frac{1}{2}}(t) \qquad (6.43)$$

This is also a formal series, containing both positive and negative powers of D. Let $(L(t))_{+}^{\frac{2m+1}{2}}$ denote the part of the formal series where the degree of the differential operator is greater than or equal to zero. The complement denoted by $(L(t))_{-}^{\frac{2m+1}{2}}$ consists of terms with only negative powers of D and we have

$$(L(t))^{\frac{2m+1}{2}} = (L(t))_{+}^{\frac{2m+1}{2}} + (L(t))_{-}^{\frac{2m+1}{2}} \qquad (6.44)$$

Let us note from Eq. (6.43) that since the highest derivative contained in $L^m(t)$ is D^{2m}, to know $(L(t))_{+}^{\frac{2m+1}{2}}$ accurately, we must know the coefficient of the D^{-2m} term in the expansion of $L^{\frac{1}{2}}(t)$. Let us also note that

$$\left[(L(t))^{\frac{2m+1}{2}} , L(t)\right] = 0 \qquad (6.45)$$

Using Eq. (6.44), we obtain

$$\left[(L(t))_{+}^{\frac{2m+1}{2}} , L(t)\right] = -\left[(L(t))_{-}^{\frac{2m+1}{2}} , L(t)\right] \qquad (6.46)$$

The left hand side is a derivative operator of degree ≥ 0 whereas the right hand side is a derivative operator of degree ≤ 0. Thus the equality would hold only if both sides correspond to a multiplicative operator. Thus we see that

$$\left[(L(t))_+^{\frac{2m+1}{2}} , L(t) \right] \qquad \text{for} \qquad m = 0,1,2.... \quad (6.47)$$

is a multiplicative operator and hence $(L(t))_+^{\frac{2m+1}{2}}$ can be identified, up to a multiplicative constant, with $B_m(t)$. Namely,

$$B_m(t) = \alpha_m (L(t))_+^{\frac{2m+1}{2}} \qquad\qquad (6.48)$$

Examples:

Following Eq. (6.48), we write

$$B_0(t) = \alpha_0 (L(t))_+^{\frac{1}{2}} \qquad\qquad (6.49)$$

Let us recall from Eq. (6.42) that

$$(L(t))^{\frac{1}{2}} = D + a_0(u) + \sum_{n=1}^{\infty} a_n(u) D^{-n}$$

so that

$$(L(t))^{\frac{1}{2}}_+ = D + a_0(u) \tag{6.50}$$

To determine $a_0(u)$, we must expand $(L(t))^{\frac{1}{2}}$ up to the D^{-1} terms. That is, let

$$(L(t))^{\frac{1}{2}} = D + a_0(u) + a_1(u)D^{-1} \tag{6.51}$$

Squaring this and keeping terms only up to D^0 we have

$$(D+a_0(u) + a_1(u)D^{-1})(D+a_0(u) + a_1(u)D^{-1}) = D^2 + \frac{1}{6}u$$

or, $D^2 + 2a_0(u)D + \left((Da_0(u)) + (a_0(u))^2\right.$

$$\left. + 2a_1(u)\right) = D^2 + \frac{1}{6}u \tag{6.52}$$

Comparing terms on both sides, we obtain

$$a_0(u) = 0$$

$$a_1(u) = \frac{1}{12}u \tag{6.53}$$

so that up to this order

$$(L(t))^{\frac{1}{2}} = D + \frac{1}{12} uD^{-1} \tag{6.54}$$

Note that this could have been obtained from a formal Taylor series expansion as well. For example,

$$(L(t))^{\frac{1}{2}} = (D^2 + \frac{1}{6} u)^{\frac{1}{2}}$$

$$= (1 + \frac{1}{6} uD^{-2})^{\frac{1}{2}} D$$

$$= (1 + \frac{1}{12} uD^{-2} + \ldots)D$$

$$= D + \frac{1}{12} uD^{-1} + \ldots \tag{6.55}$$

In any case, we see that

$$(L(t))^{\frac{1}{2}}_{+} = D \tag{6.56}$$

Thus we recognize that with $\alpha_0{=}1$, this is the same $B(t)$ which we had found in Eq. (6.29). Namely,

$$B_0(t) = (L(t))^{\frac{1}{2}}_{+} = D \tag{6.57}$$

If we expand $(L(t))^{\frac{1}{2}}$ consistently to the next term, we obtain

$$(L(t))^{\frac{1}{2}} = D + \frac{1}{12} uD^{-1} - \frac{1}{24} u_x D^{-2}$$

Thus

$$(L(t))^{\frac{3}{2}} = (D^2 + \frac{1}{6} u)(L(t))^{\frac{1}{2}}$$

$$= D^3 + \frac{1}{12} D^2 u D^{-1} - \frac{1}{24} D^2 u_x D^{-2} + \frac{1}{6} uD$$

$$+ \frac{1}{72} u^2 D^{-1} - \frac{1}{144} u u_x D^{-2}$$

$$= D^3 + \frac{1}{8} u_x + \frac{1}{4} uD + 0(D^{-1})$$

$$\text{or,} \quad (L(t))^{\frac{3}{2}} = D^3 + \frac{1}{8} (Du+uD) + 0(D^{-1}) \tag{6.58}$$

Consequently,

$$(L(t))_+^{\frac{3}{2}} = D^3 + \frac{1}{8} (Du+uD) \tag{6.59}$$

With $\alpha_1 = 4$, we recognize that

$$B_1(t) = \alpha_1 (L(t))^{\frac{3}{2}} = 4D^3 + \frac{1}{2} (Du+uD) \tag{6.60}$$

is the same B(t) we had determined earlier in Eq. (6.37) for the KdV equation.   All   the $B_m(t)$'s can likewise be determined up to a multiplicative constant.

Lenard's Derivation of the KdV Equation:

To understand further the interplay between the linear Schrödinger equation and the entire hierarchy of the KdV equations, let us ask if we can derive the latter from the former. The crucial assumption in this derivation clearly would be the t independence of the spectral parameter λ.

The Schrödinger equation is

$$\frac{\partial^2 \psi(x,t)}{\partial x^2} + (\tfrac{1}{6} u(x,t)+\lambda)\psi(x,t) = 0$$

or, $\quad \psi_{xx} + (\tfrac{1}{6} u+\lambda)\psi = 0 \qquad\qquad\qquad (6.61)$

Differentiating with respect to t we obtain

$$\psi_{xxt} + (\tfrac{1}{6} u+\lambda)\psi_t + \tfrac{1}{6} u_t\psi = 0 \qquad\qquad (6.62)$$

Let us choose $\psi(x,t)$ to be normalized to unity for all t so that

$$\int_{-\infty}^{\infty} dx\psi^2(x,t) = 1 \qquad\qquad\qquad\qquad (6.63)$$

Consequently, if we multiply Eq. (6.62) with $\psi(x,t)$ and integrate over the whole x-axis, we obtain

$$\int_{-\infty}^{\infty} dx\left(\psi\psi_{xxt} + \tfrac{1}{6} u\psi\psi_t + \tfrac{1}{6} u_t\psi^2\right) = 0 \qquad (6.64)$$

On the other hand, we see from Eq. (6.61)   that we can write

$$\lambda = -\int_{-\infty}^{\infty} dx \left(\psi\psi_{xx} + \frac{1}{6}\, u\psi^2\right) \tag{6.65}$$

and since λ is assumed to be independent of t, we obtain

$$\lambda_t = 0 = -\int_{-\infty}^{\infty} dx \left(\psi_t\psi_{xx} + \psi\psi_{xxt} + \frac{1}{6}\, u_t\psi^2 + \frac{1}{3}\, u\psi\psi_t\right)$$

$$= -\int_{-\infty}^{\infty} dx \left(2\psi\psi_{xxt} + \frac{1}{6}\, u_t\psi^2 + \frac{1}{3}\, u\psi\psi_t\right) \tag{6.66}$$

Using relation (6.64), the above expression becomes

$$\int_{-\infty}^{\infty} dx \left(\frac{1}{3}\, u\psi\psi_t + \frac{1}{3}\, u_t\psi^2 - \frac{1}{6}\, u_t\psi^2 - \frac{1}{3}\, u\psi\psi_t\right) = 0$$

or, $$\int_{-\infty}^{\infty} dx u_t\psi^2 = 0 \tag{6.67}$$

Therefore, the form of u_t must be such that the integrand in Eq. (6.67) is a total divergence. The most general form for this is

$$u_t\psi^2 = \frac{\partial}{\partial x} \left(A(u)\psi_x^2 + B(u)\psi\psi_x + C(u)\psi^2\right) \tag{6.68}$$

Here A(u), B(u) and C(u) are functionals of u and depend on

λ.   Let us emphasize that this is the most general form of the integrand since any higher derivative of ψ can be reduced using the Schrödinger equation.   Writing out the terms explicitly we have

$$u_t\psi^2 = A_x\psi_x^2 + 2A\psi_x\psi_{xx} + B_x\psi\psi_x + B(\psi_x)^2$$

$$+ B\psi\psi_{xx} + C_x\psi^2 + 2C\psi\psi_x$$

or, $u_t\psi^2 = (A_x+B)(\psi_x)^2 + \left(B_x + 2C - 2A(\tfrac{1}{6}\,u+\lambda)\right)\psi\psi_x$

$$+ \left(C_x - B\,(\tfrac{1}{6}\,u+\lambda)\right)\psi^2 \qquad (6.69)$$

Comparing both sides we obtain

$$A_x + B = 0$$

or, $B = -A_x$

$$(6.70)$$

$$B_x + 2C - 2A(\tfrac{1}{6}\,u+\lambda) = 0$$

or, $C = A(\tfrac{1}{6}\,u+\lambda) + \tfrac{1}{2}\,A_{xx}$

and

$$u_t = C_x - B(\tfrac{1}{6}\,u+\lambda)$$

$$= \tfrac{1}{2}\,A_{xxx} + 2A_x\,(\tfrac{1}{6}\,u+\lambda) + \tfrac{1}{6}\,u_x A$$

or, $u_t = \tfrac{1}{2}\,(D^3 + \tfrac{1}{3}\,(Du+uD))A + 2\lambda DA \qquad (6.71)$

Since u_t is independent of λ, we must choose A(u) to be a function of λ such that the above relation is consistent. Expanding as a power series in (-4λ) (we choose this to be consistent with our earlier convention of Eq. (3.58))

$$A(u) = 2 \sum_{j=0}^{n} A_j(u)(-4\lambda)^{n-j} \qquad (6.72)$$

and substituting into Eq. (6.71), we obtain the recursion relation

$$(D^3 + \frac{1}{3}(Du+uD))A_j = DA_{j+1} \qquad j = 0,1,\ldots n-1 \quad (6.73)$$

$$A_0 = 1$$

and

$$u_t = (D^3 + \frac{1}{3}(Du+uD))A_n \qquad (6.74)$$

We recognize that the A_j's satisfy the same recursion relation (see Eq. (3.58)) as the conserved quantities of the KdV equation. Thus we can identify

$$A_j = \frac{\delta H_j}{\delta u(x)} \qquad (6.75)$$

We have, therefore, recovered the functional recursion relations of the KdV equation as well as all the equations

of the KdV hierarchy since each equation simply corresponds
to a particular form of A where the coefficient functions
are completely determined.   This shows that the entire
hierarchy of equations can be obtained from the same
Schrödinger equation if we assume the spectral parameter to
be t-independent.

References:

Gel'fand, I. M. and L. A. Dikii, Russ. Math. Surveys $\underline{30}$,
        77 (1975).

Kruskal, M. in Dynamical Systems, Theory and Applica-
        tions, Ed. J. Moser, Springer-Verlag, 1974.

Lax, P. D., Comm. Pure Appl. Math. $\underline{21}$, 467 (1968).

Lax, P. D., Comm. Pure Appl. Math. $\underline{28}$, 141 (1975).

Lenard, A. (unpublished) as reported in C. S. Gardner,
        J. M. Greene, M. D. Kruskal and R. M. Miura,
        Comm. Pure Appl. Math. $\underline{27}$, 97 (1974).

CHAPTER 7

MORE ON KdV

In this chapter, we will discuss various other fascinating aspects of the KdV equation. First of all we will
try to clarify the physical meaning of the spectral parameter of the Schrödinger equation. We will show how the
conserved quantities can be seen to be in involution in the
Lax method. Finally, we will bring out the symmetry group
structure associated with the KdV equation.

The Spectral Parameter:

Let us next understand the physical significance of the
conserved eigenvalues of the Schrödinger equation in the
case of the KdV equation. To do that let us denote the KdV
equation compactly as

$$u_t = uu_x + u_{xxx} = K(u) \qquad (7.1)$$

Furthermore, we can construct a one parameter family of
solutions, $u^{(\epsilon)}(x,t)$, of the KdV equation in the following
way. Let

$$u^{(\epsilon)}(x,0) = u(x,0) + \epsilon f(x) \qquad (7.2)$$

where $f(x)$ is a smooth function vanishing asymptotically.

We denote the solution of the KdV equation corresponding to this initial data by $u^{(\epsilon)}(x,t)$. This will in general be a power series in ϵ of the form

$$u^{(\epsilon)}(x,t) = u(x,t) + \epsilon v(x,t) + 0(\epsilon^2)$$

where   (7.3)

$$v(x,t) = \frac{du^{(\epsilon)}(x,t)}{d\epsilon}\Bigg|_{\epsilon=0}$$

(Here $v(x,t)$ is not the MKdV variable.) Furthermore, let us define

$$\frac{dK(u^{(\epsilon)})}{d\epsilon}\Bigg|_{\epsilon=0} = \frac{dK(u+\epsilon v)}{d\epsilon}\Bigg|_{\epsilon=0} = M(u)v \qquad (7.4)$$

Clearly this is a linear functional of v and is known as the Frechet derivative of $K(u)$. $M(u)$ is more commonly known as the functional derivative of $K(u)$. Let me emphasize that $M(u)$ would in general be an operator involving derivatives. Let us also note from Eq. (7.1) that $u^{(\epsilon)}$ would satisfy the equation

$$u_t^{(\epsilon)} = K(u^{(\epsilon)}) \qquad (7.5)$$

so that differentiating both sides with respect to ϵ and setting $\epsilon=0$ we obtain the linear equation

$$v_t = M(u)v \qquad (7.6)$$

Let us also, for simplicity, introduce the following notation for the functional derivatives of the conserved quantities. If $H_n(u)$ is a conserved quantity, then its functional derivative is obtained from

$$\left.\frac{dH_n(u^{(\epsilon)})}{d\epsilon}\right|_{\epsilon=0} = \left.\frac{dH_n(u+\epsilon v)}{d\epsilon}\right|_{\epsilon=0} = (G_n(u),v)$$

$$= \int_{-\infty}^{\infty} dx \ G_n(u(x,t))v(x,t) \qquad (7.7)$$

so that

$$G_n(u(x,t)) = \frac{\delta H_n(u)}{\delta u(x,t)} \qquad (7.8)$$

We also readily identify from Eqs. (3.12) and (7.1)

$$K(u) = \frac{\partial G_2(u(x,t))}{\partial x} = \frac{\partial}{\partial x}\frac{\delta H_2}{\delta u(x,t)} \qquad (7.9)$$

Let us note here that since $H_n(u^{(\epsilon)})$ is conserved for all values of ϵ,

$$(G_n(u(x,t)) \ , \ v(x,t)) \qquad (7.10)$$

must be time independent. Taking the time derivative of Eq. (7.10), we obtain

$$\left(\frac{\partial G_n}{\partial t} \, , \, v\right) + \left(G_n \, , \, \frac{\partial v}{\partial t}\right) = 0 \tag{7.11}$$

Furthermore, using the form of the evolution for v from Eq. (7.6) we obtain

$$\left(\frac{\partial G_n}{\partial t} \, , \, v\right) + \left(G_n \, , \, M(u)v\right) = 0$$

or, $\quad \left((\frac{\partial}{\partial t} + M^+(u))G_n \, , \, v\right) = 0 \tag{7.12}$

Here $M^+(u)$ is the adjoint of $M(u)$ with respect to the inner product (,).

Since v can be prescribed arbitrarily by choosing an appropriate initial configuration, we conclude that the above relation would be true only if

$$\left(\frac{\partial}{\partial t} + M^+(u)\right)G_n(u(x,t)) = 0 \tag{7.13}$$

That is, the functional derivative of each of the conserved quantities must satisfy the above equation. In particular, since

$$H_1 = \frac{1}{2} \int_{-\infty}^{\infty} dx \, u^2(x,t)$$

we see that

$$G_1 = \frac{\delta H_1}{\delta u(x,t)} = u(x,t) \tag{7.14}$$

and must satisfy

$$\left(\frac{\partial}{\partial t} + M^+(u)\right)G_1 = \left(\frac{\partial}{\partial t} + M^+(u)\right)u = 0 \qquad (7.15)$$

Comparing this with the KdV equation, Eq. (7.1) we see that

$$K(u) = -M^+(u)u \qquad (7.16)$$

This, of course, can be directly checked.

$$K(u) = uu_x + u_{xxx}$$

Hence

$$\left.\frac{dK(u+\epsilon v)}{d\epsilon}\right|_{\epsilon=0} = M(u)v$$

gives

$$M(u) = \frac{\partial^3}{\partial x^3} + u\frac{\partial}{\partial x} + u_x \qquad (7.17)$$

It is obvious, therefore, that

$$M^+(u) = -\frac{\partial^3}{\partial x^3} - u\frac{\partial}{\partial x} \qquad (7.18)$$

which directly leads to

$$K(u) = -M^+(u)u \qquad\qquad (7.19)$$

It is also useful to note from Eq. (7.18) the operator nature of $M(u)$.

Let us now consider the case where the solution of the KdV equation represents a solitary wave. That is, let

$$u(x,t) = s(x+ct) \qquad\qquad (7.20)$$

where c is the speed of the solitary wave which moves to the left. This wave satisfies the equation

$$\frac{\partial s(x+ct)}{\partial t} = K(s) = -M^+(s)s$$

$$\text{or,} \quad c\,\frac{\partial s}{\partial x} = -M^+(s)s$$

$$\qquad\qquad (7.21)$$

$$\text{or,} \quad \left(c\,\frac{\partial}{\partial x} + M^+(s)\right)s = 0$$

On the other hand, the functional derivatives of the conserved quantities also satisfy the same equation in the present case. Namely,

$$\left(\frac{\partial}{\partial t} + M^+(s)\right)G_n(s(x+ct)) = 0$$

$$\text{or,} \quad \left(c\,\frac{\partial}{\partial x} + M^+(s)\right)G_n(s) = 0 \qquad\qquad (7.22)$$

Comparing Eqs. (7.21) and (7.22), it can be shown that

$$G_n(s) \propto s \qquad (7.23)$$

That is, the functional derivatives of all the conserved quantities for the case of solitary waves are indeed proportional to the solitary waves. Another way of expressing this fact is to note that for such solutions, all the conserved quantities must take the same form given by

$$H_n(s) \propto \frac{1}{2} \int_{-\infty}^{\infty} dx\ s^2(x+ct) \qquad (7.24)$$

Let us now turn to the Schrödinger equation for the one parameter family of potentials. Namely,

$$\psi_{xx}^{(\epsilon)} + \left(\frac{1}{6} u^{(\epsilon)} + \lambda^{(\epsilon)}\right)\psi^{(\epsilon)} = 0 \qquad (7.25)$$

where

$$u^{(\epsilon)} = u + \epsilon v + 0(\epsilon^2)$$

$$\psi^{(\epsilon)} = \psi + \epsilon\phi + 0(\epsilon^2) \qquad (7.26)$$

$$\lambda^{(\epsilon)} = \lambda + \epsilon \left.\frac{d\lambda}{d\epsilon}\right|_{\epsilon=0} + 0(\epsilon^2)$$

so that for $\epsilon=0$ we recover the usual equation

$$\psi_{xx} + (\frac{1}{6} u + \lambda)\psi = 0$$

Now taking the derivative of Eq. (7.25) with respect to ϵ and setting $\epsilon=0$ we obtain

$$\phi_{xx} + (\tfrac{1}{6} v + \tfrac{d\lambda}{d\epsilon}\Big|_{\epsilon=0})\psi + (\tfrac{1}{6} u + \lambda)\phi = 0 \qquad (7.27)$$

Taking the inner product of Eq. (7.27) with ψ and using the fact that ψ satisfies the usual Schrödinger equation of (6.61) we obtain

$$\left(\psi \; , \; \phi_{xx} + (\tfrac{1}{6} u + \lambda)\phi + (\tfrac{1}{6} v + \tfrac{d\lambda}{d\epsilon}\Big|_{\epsilon=0})\psi\right) = 0$$

or, $\left(\psi_{xx} + (\tfrac{1}{6} u + \lambda)\psi \; , \; \phi\right) + \tfrac{d\lambda}{d\epsilon}\Big|_{\epsilon=0} (\psi,\psi) + \tfrac{1}{6} (\psi, v\psi) = 0$

or, $\tfrac{d\lambda}{d\epsilon}\Big|_{\epsilon=0} = -\tfrac{1}{6} (\psi \; , \; v\psi) = -\tfrac{1}{6} (\psi^2, v) \qquad (7.28)$

Here we are assuming that ψ is normalized to unity. Furthermore, from the definition of the Frechet derivative in Eq. (7.7) and (7.8), we conclude that

$$\frac{\delta\lambda}{\delta u(x,t)} = G_\lambda(u) = -\tfrac{1}{6} \psi^2(u) \qquad (7.29)$$

We also note, following our discussion in Eq. (7.23), that since $\lambda(u)$ is conserved under the KdV flow, for a solitary wave solution

$$G_\lambda(s) = -\tfrac{1}{6} \psi^2(s) \propto s$$

or, $\psi(s) \propto s^{1/2}$ (7.30)

That is, in the case of a solitary wave potential, the eigenfunction of the Schrödinger equation is proportional to the square root of the solitary wave solution. We have, of course, already seen this for the one soliton potential. Namely, recall from Eqs. (4.32) and (4.35) that when

$$u(x,0) = 12\text{sech}^2 x$$

the bound state wave function is given by

$$\psi(x) = \tfrac{1}{2} \text{ sech } x$$

However, let us next see the consequences of the above identification. Putting the form of the wave function from Eq. (7.30) back into the Schrödinger equation we obtain

$$\left(\frac{\partial^2}{\partial x^2} + \left(\tfrac{1}{6} s + \lambda\right)\right) s^{1/2} = 0$$

or, $s_{xx} - \tfrac{1}{2} s^{-1} s_x^2 + \tfrac{1}{3} s^2 + 2\lambda s = 0$ (7.31)

Differentiating the above equation with respect to x, we obtain

$$s_{xxx} - s^{-1} s_x \left(s_{xx} - \tfrac{1}{2} s^{-1} s_x^2 + \tfrac{1}{3} s^2 + 2\lambda s\right) + s s_x + 4\lambda s_x = 0$$

or, $s_{xxx} + ss_x + 4\lambda s_x = 0$ (7.32)

We have used Eq. (7.31) to obtain the form of the last equation. Furthermore, using the form of $M^+(s)$ from Eq. (7.18), we can write the above equation as

$$\left(4\lambda \frac{\partial}{\partial x} - M^+(s)\right)s = 0$$ (7.33)

On the other hand, since s represents a solitary wave, it must satisfy Eq. (7.21). Comparing then, we obtain

$$c = -4\lambda(s)$$ (7.34)

This gives a physical meaning to the spectral parameter of the Schrödinger equation. Namely, the discrete eigenvalues correspond to the speeds of the solitary waves up to the same multiplicative constant of $-1/4$. In fact, let us recall the results from our one soliton calculation of Chapter 4. Namely, for (see Eqs. (4.32), (4.34) and (4.49))

$$u(x,0) = 12\text{sech}^2 x$$

$$\lambda = -1$$

the inverse scattering method yielded the complete solution to be

$$u(x,t) = 12\text{sech}^2(x+4t)$$

so that

$$c = 4 = -4\lambda \qquad\qquad (7.35)$$

This new meaning of the eigenvalues leads to an intuitive feeling for why they should be conserved. Let us recall that the solitary waves preserve their shape. Furthermore, as we have seen in Eq. (2.24), their shape is intimately dependent on the speed. Consequently, the speed for a solitary wave must be conserved. It is also clear that since the solitary waves can have infinitely many distinct speeds, the number of conserved quantities for the KdV equation should likewise be infinite.

Involution of the Conserved Quantities:

To see further that the eigenvalues of the Schrödinger equation, which are conserved, are also in involution, let us note that if

$$\psi_{xx} + (\tfrac{1}{6} u + \lambda)\psi = 0$$

then, as we have derived earlier in Eq. (7.29)

$$\frac{\delta\lambda}{\delta u(x,t)} = -\tfrac{1}{6} \psi^2$$

Let us also note that

$$D(\psi^2) = 2\psi\psi_x$$

$$D^2(\psi^2) = D(2\psi\psi_x) = 2(\psi_x)^2 + 2\psi\psi_{xx}$$

$$= 2(\psi_x)^2 - (\tfrac{1}{3}\,u+2\lambda)\psi^2$$

(7.36)

$$D^3(\psi^2) = D(2(\psi_x)^2 - (\tfrac{1}{3}\,u+2\lambda)\psi^2)$$

$$= 4\psi_x\psi_{xx} - \tfrac{1}{3}\,u_x\psi^2 - 2(\tfrac{1}{3}\,u+2\lambda)\psi_x\psi$$

$$= -4(\tfrac{1}{6}\,u+\lambda)\psi_x\psi - \tfrac{1}{3}\,u_x\psi^2 - 2(\tfrac{1}{3}\,u+2\lambda)\psi_x\psi$$

$$= -(\tfrac{4}{3}\,u+8\lambda)\psi_x\psi - \tfrac{1}{3}\,u_x\psi^2$$

Consequently, it follows that

$$\left(D^3 + \tfrac{1}{3}\,(Du+uD)\right)\psi^2$$

$$= -(\tfrac{4}{3}\,u+8\lambda)\psi_x\psi - \tfrac{1}{3}\,u_x\psi^2 + \tfrac{1}{3}\,u_x\psi^2 + \tfrac{2}{3}\,u(2\psi\psi_x)$$

$$= -8\lambda\psi_x\psi = -4\lambda D(\psi^2)$$

(7.37)

Thus we see that

$$\left(D^3 + \tfrac{1}{3}\,(Du+uD)\right)\frac{\delta\lambda}{\delta u(x,t)} = -4\lambda D\,\frac{\delta\lambda}{\delta u(x,t)}$$

(7.38)

Hence it follows that the Poisson bracket between the λ's can be calculated and it satisfies

$$\{\lambda_i, \lambda_j\} = \int_{-\infty}^{\infty} dx \, \frac{\delta\lambda_i}{\delta u(x,t)} \, D \, \frac{\delta\lambda_j}{\delta u(x,t)}$$

$$= -\frac{1}{4\lambda_j} \int_{-\infty}^{\infty} dx \, \frac{\delta\lambda_i}{\delta u(x,t)} \left(D^3 + \frac{1}{3} (Du+uD) \right) \frac{\delta\lambda_j}{\delta u(x,t)}$$

$$= \frac{1}{4\lambda_j} \int_{-\infty}^{\infty} dx \, \left(D^3 + \frac{1}{3} (Du+uD) \right) \frac{\delta\lambda_i}{\delta u(x,t)} \, \frac{\delta\lambda_j}{\delta u(x,t)}$$

$$= -\frac{\lambda_i}{\lambda_j} \int_{-\infty}^{\infty} dx \, D \, \frac{\delta\lambda_i}{\delta u(x,t)} \, \frac{\delta\lambda_j}{\delta u(x,t)}$$

$$= \frac{\lambda_i}{\lambda_j} \int_{-\infty}^{\infty} dx \, \frac{\delta\lambda_i}{\delta u(x,t)} \, D \, \frac{\delta\lambda_j}{\delta u(x,t)}$$

or, $$\{\lambda_i, \lambda_j\} = \frac{\lambda_i}{\lambda_j} \{\lambda_i, \lambda_j\} \qquad\qquad (7.39)$$

Therefore, if $\lambda_i \neq \lambda_j$

$$\{\lambda_i, \lambda_j\} = 0 \qquad\qquad (7.40)$$

If $\lambda_i = \lambda_j$, then the Poisson bracket, of course, vanishes by antisymmetry. This, therefore, shows that the infinite number of conserved quantities are also in involution and consequently, the KdV equation is integrable. Indeed, if we can find a Lax pair to represent a given equation, the equation would be integrable.

KdV Equation and the Group SL(2;R):

In this section, we will try to bring out another fascinating aspect of the KdV equation. Namely, we will derive the KdV equation from the structure equation of the Lie Group SL(2;R). Besides establishing the interesting connection between the integrable models and the symmetry groups, this discussion would provide a natural motivation for our later discussion of the method of zero curvature condition as well as the AKNS (Ablowitz-Kaup-Newell-Segur) method. However, to avoid any possible confusion, let us emphasize here that this derivation claims no direct connection between SL(2;R) and the possible symmetry group associated with the infinite number of conserved charges.

Let us recall that the Lie Group SL(2;R) is defined by the group properties of the set of all 2×2 real matrices with determinant unity. The Lie algebra of the group consists of three hermitian generators T_a, with a = 1,2,3, which in a given basis satisfy the commutation relations

$$[T_a, T_b] = iC^c_{ab} T_c \qquad (7.41)$$

where the structure constants, C^c_{ab} , take the values

$$C^1_{23} = -C^1_{32} = -1$$

$$C^2_{12} = -C^2_{21} = -2 \qquad (7.42)$$

$$C^3_{31} = -C^3_{13} = -2$$

We parenthetically remark here, for the interested reader,
that the generators of the SL(2;R) subalgebra of the
Virasoro algebra can be identified with

$$L_0 = \frac{1}{2} T_1$$

$$L_1 = T_2 \qquad\qquad\qquad\qquad (7.43)$$

$$L_{-1} = -T_3$$

so that the product rule

$$[L_m, L_n] = i(m-n)L_{m+n} \qquad\qquad m,n = 0,\pm1 \qquad (7.44)$$

follows from Eqs. (7.41) and (7.42).

An element of a Lie group can be written in terms of the
generators of the Lie algebra as

$$g = \exp\left(i\theta^a T_a\right) \qquad\qquad\qquad (7.45)$$

In particular if the parameters θ^a are functions of space
and time, then the corresponding group elements would also
be space-time dependent.   Since we are interested in 1+1
dimensional theories, we would assume a typical element,
$g(x,t)\in$ SL(2;R), to be a space-time dependent matrix with
determinant unity which can be represented in terms of the
generators discussed earlier as

$$g(x,t) = \exp\left(i\theta^a(x,t)T_a\right) \qquad (7.46)$$

Note that $g(x,t)$ would be a matrix whose dimensionality would depend on the dimensionality of the representation.

Given a group element $g(x,t)$ we can construct a 1+1 dimensional matrix valued covariant vector A_μ as

$$A_\mu = g^{-1}(x,t)\partial_\mu g(x,t) \qquad \mu = 0,1 \qquad (7.47)$$

where

$$x^0 = t \qquad x^1 = x$$

$$\qquad (7.48)$$

$$\partial_0 = \frac{\partial}{\partial t} \qquad \partial_1 = \frac{\partial}{\partial x}$$

From the structure of A_μ in Eq. (7.47), we see that it must satisfy the equation

$$\partial_\mu A_\nu - \partial_\nu A_\mu + [A_\mu, A_\nu] = 0 \qquad (7.49)$$

This is known as the Cartan-Maurer equation or the structure equation since it can determine the structure constants of the group. In the language of gauge theories, we see that if we view A_μ as a gauge potential or a connection, then the Cartan-Maurer equation says that the field strength associated with this potential or the curvature associated with this connection vanishes. This, as we know, is synonymous with the statement that the potential is a pure gauge.

Let us note that the Cartan-Maurer connection is anti Hermitian and can be represented as

$$A_\mu(x,t) = iA_\mu^a(x,t)T_a \qquad (7.50)$$

so that the structure equation (Eq. (7.49)) takes the form

$$\partial_\mu A_\nu^a - \partial_\nu A_\mu^a - C_{bc}^a A_\mu^b A_\nu^c = 0$$

$$\mu,\nu = 0,1 \qquad a = 1,2,3 \quad (7.51)$$

These are actually a set of three equations and let us study them for a special choice of the A_μ^a variables. Namely, let

$$A_1^1 = \sqrt{-\lambda} \qquad \lambda < 0$$

$$\qquad (7.52)$$

$$A_1^3 = 6$$

Then for a=3, the structure equation gives

$$-A_{0,x}^3 + 2\left(\sqrt{-\lambda}\, A_0^3 - 6A_0^1\right) = 0$$

$$\text{or,} \quad A_0^1 = \left(\frac{\sqrt{-\lambda}}{6} A_0^3 - \frac{1}{12} A_{0,x}^3\right) \qquad (7.53)$$

Here and in what follows, a comma followed by a subscript denotes a derivative with respect to that variable.

Similarly, looking at the structure equation (7.51) for a=1, we obtain

$$-A_{0,x}^1 + \left(6A_0^2 - A_0^3 A_1^2\right) = 0$$

or, $A_0^2 = \frac{1}{6} A_{0,x}^1 + \frac{1}{6} A_0^3 A_1^2$

or, $A_0^2 = \left(\frac{\sqrt{-\lambda}}{36} A_{0,x}^3 - \frac{1}{72} A_{0,xx}^3 + \frac{1}{6} A_0^3 A_1^2\right)$ (7.54)

Finally, for a=2, we see that the structure equation leads to

$$A_{1,t}^2 - A_{0,x}^2 + 2\left(A_0^1 A_1^2 - \sqrt{-\lambda}\, A_0^2\right) = 0$$

or, $A_{1,t}^2 = A_{0,x}^2 - 2\left(A_0^1 A_1^2 - \sqrt{-\lambda}\, A_0^2\right)$

Upon using Eqs. (7.53) and (7.54), this simplifies to

$$A_{1,t}^2 = -\frac{1}{72} A_{0,xxx}^3 + \frac{1}{3} A_{0,x}^3 A_1^2 + \frac{1}{6} A_0^3 A_{1,x}^2 - \frac{\lambda}{18} A_{0,x}^3 \quad (7.55)$$

We recognize that if we further identify

$$A_1^2 = -\frac{1}{36} u(x,t)$$

                                                                    (7.56)

$$A_0^3 = A(u(x,t))$$

then equation (7.55) takes the form

$$u_t = \frac{1}{2} A_{xxx} + \frac{1}{3} A_x u + \frac{1}{6} u_x A + 2\lambda A_x$$

$$(7.57)$$

$$\text{or,} \quad u_t = \frac{1}{2}\left(D^3 + \frac{1}{3}\,(Du+uD)\right)A + 2\lambda DA$$

This is precisely the equation we had obtained, namely, Eq. (6.71), in Lenard's derivation of the KdV equation. Proceeding exactly as before then, we not only can derive the whole hierarchy of the KdV equations but also the functional recursion relation between the conserved quantities. However, the present derivation relates all of this to the structure equation for the Lie Group SL(2;R).

Let us note next that the MKdV equation can also be obtained from the same Cartan-Maurer equation in the following way. Let us choose

$$A_1^2 = A_1^3 = \frac{i}{\sqrt{6}}\,v$$

$$(7.58)$$

$$A_1^1 = \sqrt{-\lambda}$$

Here $v(x,t)$ is the dynamical variable. Then the structure equation, Eq. (7.51), for a=1 gives

$$-A_{0,x}^1 + \frac{i}{\sqrt{6}}\,v\left(A_0^2 - A_0^3\right) = 0$$

$$\text{or,} \quad \left(A_0^2 - A_0^3\right) = -i\,\sqrt{6}\left(\frac{A_{0,x}^1}{v}\right) \qquad (7.59)$$

The structure equation for a=2 gives

$$\frac{i}{\sqrt{6}} v_t - A_{0,x}^2 + 2\left(\frac{i}{\sqrt{6}} vA_0^1 - \sqrt{-\lambda} A_0^2\right) = 0$$

$$\text{or,} \quad \frac{i}{\sqrt{6}} v_t = A_{0,x}^2 - 2\left(\frac{i}{\sqrt{6}} vA_0^1 - \sqrt{-\lambda} A_0^2\right) \qquad (7.60)$$

Similarly, the structure equation for a=3 leads to

$$\frac{i}{\sqrt{6}} v_t - A_{0,x}^3 + 2\left(\sqrt{-\lambda} A_0^3 - \frac{i}{\sqrt{6}} vA_0^1\right) = 0$$

$$\text{or,} \quad \frac{i}{\sqrt{6}} v_t = A_{0,x}^3 - 2\left(\sqrt{-\lambda} A_0^3 - \frac{i}{\sqrt{6}} vA_0^1\right) \qquad (7.61)$$

Subtracting Eq. (7.61) from Eq. (7.60), we obtain

$$\left(A_0^2 - A_0^3\right)_x - \frac{4i}{\sqrt{6}} vA_0^1 + 2\sqrt{-\lambda} \left(A_0^2 + A_0^3\right) = 0$$

$$\text{or,} \quad \left(A_0^2 + A_0^3\right) = \frac{2i}{\sqrt{-6\lambda}} vA_0^1 - \frac{1}{2\sqrt{-\lambda}} \left(A_0^2 - A_0^3\right)_x \qquad (7.62)$$

Using Eq. (7.59), we then obtain from the above relation

$$\left(A_0^2 + A_0^3\right)_x = \frac{2i}{\sqrt{-6\lambda}} \left(vA_0^1\right)_x + \frac{i\sqrt{6}}{2\sqrt{-\lambda}} \left(\frac{A_{0,x}^1}{v}\right)_{xx} \qquad (7.63)$$

Similarly adding Eqs. (7.60) and (7.61) we get

$$\frac{2i}{\sqrt{6}} v_t = \left(A_0^2 + A_0^3\right)_x + 2\sqrt{-\lambda} \left(A_0^2 - A_0^3\right)$$

$$= \frac{2i}{\sqrt{-6\lambda}} \left(vA_0^1\right)_x + \frac{i\sqrt{6}}{2\sqrt{-\lambda}} \left(\frac{A_{0,x}^1}{v}\right)_{xx}$$

$$- 2i \sqrt{-6\lambda} \left(\frac{A_{0,x}^1}{v}\right)$$

or, $\quad v_t = \frac{1}{\sqrt{-\lambda}} \left(vA_0^1\right)_x + \frac{3}{2\sqrt{-\lambda}} \left(\frac{A_{0,x}^1}{v}\right)_{xx}$

$$- 6\sqrt{-\lambda} \left(\frac{A_{0,x}^1}{v}\right) \qquad (7.64)$$

If we now let

$$A_0^1 = \sqrt{-\lambda} \left(-4\lambda + \frac{1}{3} v^2\right) \qquad (7.65)$$

then Eq. (7.64) takes the form

$$v_t = v^2 v_x + v_{xxx} \qquad (7.66)$$

This is , of course, the MKdV equation and is also obtained
from the structure equation for the Group SL(2;R). This is
not surprising since the KdV and the MKdV equations are
related and hence can be traced to the same symmetry group.

References:

Chern, S-S. and C-K. Peng, Manuscripta Mathematica 28,
        207 (1979).

Gardner, C. S., J. M. Greene, M. D. Kruskal and R. M.
        Miura, Comm. Pure Appl. Math. 27, 97 (1974).

Kruskal, M. in Dynamical Systems, Theory and Applica-
        tions, Ed. J. Moser, Springer-Verlag, 1974.

Lax, P. D., Comm. Pure Appl. Math. 21, 467 (1968).

Lax, P. D., Comm. Pure Appl. Math. 28, 141 (1975).

CHAPTER 8

MULTI-SOLITON SOLUTIONS

In the earlier chapters, we showed that the KdV equation possesses soliton solutions. We discussed the general properties of soliton solutions and explicitly constructed the one soliton solution. In this chapter, we will describe how to construct the multi-soliton solutions of the KdV equation. This is normally done through the application of Bäcklund transformations which we describe next.

Bäcklund Transformations:

Bäcklund transformations originated in the study of surfaces of constant negative curvature. Roughly speaking they can be described as follows. Given a higher order differential equation in the variable u(x,t), namely,

$$P(u(x,t)) = 0 \qquad\qquad (8.1)$$

a Bäcklund transformation is a transformation to a new variable v(x,t) defined by a pair of first order equations

$$\frac{\partial u}{\partial x} = f(u(x,t),v(x,t))$$

$$\qquad\qquad (8.2)$$

$$\frac{\partial u}{\partial t} = g(u(x,t),v(x,t))$$

where f and g depend on u, v and their derivatives in such a way that the higher order equation, namely, Eq. (8.1), arises as the integrability condition of the two first order equations. We will discuss Bäcklund transformations with examples. But let us just note here that a Bäcklund transformation may relate the solution of the original equation to that of another which is easier to solve. It may also relate one solution of the given equation to another which we may already know. It is in connection with the latter that the Bäcklund transformations first appeared and it is in that spirit that we will use Bäcklund transformations to construct multi-soliton solutions of the KdV equation.

Examples:

i)   **Liouville Equation:**

Let us consider the Liouville equation in 1+1 dimension. This is described as

$$\left(\frac{\partial^2}{\partial t^2} - \frac{\partial^2}{\partial x^2}\right) u(x,t) = e^{u(x,t)} \tag{8.3}$$

where $u(x,t)$ is the dynamical variable. For simplicity, let us make a transformation to the light cone variables defined by

$$x^{\pm} = t \pm x \tag{8.4}$$

In these variables, the Liouville equation takes the form

$$\partial_+\partial_- u = e^u \tag{8.5}$$

where we have defined

$$\partial_+ = \frac{\partial}{\partial x^+}$$

$$\tag{8.6}$$

$$\partial_- = \frac{\partial}{\partial x^-}$$

Let us next define the Bäcklund transformation, to the variable $v(x^+, x^-)$, by

$$\partial_+ u = -\partial_+ v + \alpha\, e^{\frac{1}{2}(u-v)} \tag{8.7}$$

$$\partial_- u = \partial_- v + \frac{2}{\alpha}\, e^{\frac{1}{2}(u+v)} \tag{8.8}$$

where α is an arbitrary constant. Taking the derivative of Eq. (8.7) with respect to x^- and that of Eq. (8.8) with respect to x^+, we obtain

$$\partial_-\partial_+ u = -\partial_-\partial_+ v + \frac{\alpha}{2}\, e^{\frac{1}{2}(u-v)} (\partial_- u - \partial_- v)$$

$$= -\partial_-\partial_+ v + e^u \tag{8.9}$$

$$\partial_+\partial_-u = \partial_+\partial_-v + \frac{1}{\alpha}\, e^{\frac{1}{2}\,(u+v)}\,(\partial_+u+\partial_+v)$$

$$= \partial_+\partial_-v + e^u \qquad\qquad (8.10)$$

Thus the integrability conditions for equations (8.7) and (8.8) become

$$\partial_+\partial_-u = e^u \qquad\qquad\qquad (8.11)$$

$$\partial_+\partial_-v = 0 \qquad\qquad\qquad (8.12)$$

We recognize Eq. (8.11) as the Liouville equation whereas Eq. (8.12) is simply the wave equation whose solutions are well known. The Bäcklund transformation in this case, therefore, connects the solutions of the Liouville equation with those of the wave equation.

To see exactly how the actual solution is obtained, let us note that the general solution of the wave equation takes the form

$$v(x^+,x^-) = f(x^+) + g(x^-) \qquad\qquad (8.13)$$

That is, it is a superposition of a left moving and a right moving wave. Substituting this form into Eq. (8.7) we obtain

$$\partial_+(u+f) = \alpha e^{\frac{1}{2}\,(u-f-g)} \qquad\qquad (8.14)$$

Furthermore, noting that $g(x^-)$ does not depend on x^+, we can rewrite the above equation as

$$e^{-\frac{1}{2}(u+f-g)}\partial_+(u+f-g) = \alpha e^{-f(x^+)} \qquad (8.15)$$

Integrating with respect to x^+ we obtain

$$e^{-\frac{1}{2}(u+f-g)} = -\frac{\alpha}{2}\int^{x^+} dx'^+ \, e^{-f(x'^+)} + a(x^-)$$

$$= \alpha P(x^+) + a(x^-) \qquad (8.16)$$

Here we have introduced a constant of integration $a(x^-)$ and defined the negative one half of the integral on the right hand side to be $P(x^+)$. Similarly, we obtain from Eq. (8.8)

$$\partial_-(u-g) = \frac{2}{\alpha} e^{\frac{1}{2}(u+f+g)}$$

$$\text{or,} \quad e^{-\frac{1}{2}(u+f-g)}\partial_-(u+f-g) = \frac{2}{\alpha} e^{-g(x^-)}$$

$$\text{or,} \quad e^{-\frac{1}{2}(u+f-g)} = -\frac{1}{\alpha}\int^{x^-} dx'^- \, e^{-g(x'^-)} + b(x^+)$$

$$= \frac{2}{\alpha} Q(x^-) + b(x^+) \qquad (8.17)$$

Comparing Eqs. (8.16) and (8.17) we see that

$$a(x^-) = \frac{2}{\alpha} Q(x^-)$$

$$(8.18)$$

$$b(x^+) = \alpha P(x^+)$$

so that

$$e^{-\frac{1}{2}(u+f-g)} = \alpha P(x^+) + \frac{2}{\alpha} Q(x^-)$$

or, $\quad -\frac{1}{2}(u+f-g) = \log\left(\alpha P(x^+) + \frac{2}{\alpha} Q(x^-)\right)$

or, $\quad u(x^+, x^-) = -f(x^+) + g(x^-)$

$$-2\log\left(\alpha P(x^+) + \frac{2}{\alpha} Q(x^-)\right) \qquad (8.19)$$

The one parameter family of Bäcklund transformations of Eqs. (8.7) and (8.8), therefore, has generated a one parameter family of solutions of the Liouville equation from the solutions of the wave equation.

ii) **Sine-Gordon Equation:**

The Sine-Gordon equation is the oldest example of the study of Bäcklund transformations.   Here the transformation relates one solution of the equation to another. This model also exemplifies some of the other properties associated with the Bäcklund transformations. The equation is given by

$$\frac{\partial^2 \omega}{\partial t^2} - \frac{\partial^2 \omega}{\partial x^2} = \sin \omega \qquad\qquad (8.20)$$

which in terms of the light cone variables, described in Eqs. (8.4) and (8.6), becomes

$$\partial_+ \partial_- \omega = \sin \omega \qquad\qquad (8.21)$$

The Bäcklund transformations, in the present case, to the variable $\omega_1(x^+, x^-)$, are defined by the pair of equations

$$\partial_+ \omega = \partial_+ \omega_1 + 2a \sin \left(\frac{\omega_1 + \omega}{2}\right) \qquad\qquad (8.22)$$

$$\partial_- \omega = -\partial_- \omega_1 - \frac{2}{a} \sin \left(\frac{\omega_1 - \omega}{2}\right) \qquad\qquad (8.23)$$

where a is a constant. Differentiating Eq. (8.22) with respect to x^-, we obtain

$$\partial_- \partial_+ \omega = \partial_- \partial_+ \omega_1 + a \cos \left(\frac{\omega_1 + \omega}{2}\right) \partial_- (\omega_1 + \omega)$$

$$= \partial_- \partial_+ \omega_1 - 2\cos \left(\frac{\omega_1 + \omega}{2}\right) \sin \left(\frac{\omega_1 - \omega}{2}\right)$$

$$\text{or,} \quad \partial_- \partial_+ \omega = \partial_- \partial_+ \omega_1 - (\sin \omega_1 - \sin \omega) \qquad\qquad (8.24)$$

Similarly, differentiating Eq. (8.23) with respect to x^+, we obtain

$$\partial_+\partial_-\omega = -\partial_+\partial_-\omega_1 - \frac{1}{a} \cos\left(\frac{\omega_1-\omega}{2}\right) \partial_+(\omega_1-\omega)$$

$$= -\partial_+\partial_-\omega_1 + 2\cos\left(\frac{\omega_1-\omega}{2}\right) \sin\left(\frac{\omega_1+\omega}{2}\right)$$

$$\text{or,} \quad \partial_+\partial_-\omega = -\partial_+\partial_-\omega_1 + (\sin\omega_1+\sin\omega) \tag{8.25}$$

The integrability conditions for Eqs. (8.22) and (8.23) then give

$$\partial_+\partial_-\omega = \sin\omega \tag{8.26}$$

$$\partial_+\partial_-\omega_1 = \sin\omega_1 \tag{8.27}$$

Thus we see that not only ω, but ω_1 also satisfies the Sine-Gordon equation so that in this case, the Bäcklund transformations relate one solution of the Sine-Gordon equation to another. This is quite useful in constructing the solutions of the Sine-Gordon equation, in particular, since we know that $\omega=0$ is a solution. In that case, the Bäcklund transformations of Eqs. (8.22) and (8.23) become

$$\partial_+\omega_1 = -2a \sin\frac{\omega_1}{2} \tag{8.28}$$

$$\partial_-\omega_1 = -\frac{2}{a} \sin\frac{\omega_1}{2} \tag{8.29}$$

Let us define new variables

$$\tilde{x}^+ = ax^+ \qquad\qquad \tilde{\partial}_+ = \frac{1}{a} \partial_+ \tag{8.30}$$

$$\tilde{x}^- = \frac{1}{a} \, x^- \qquad\qquad \tilde{\partial}_- = a\partial_- \qquad\qquad (8.31)$$

In terms of these, the equations (8.28) and (8.29) become

$$\tilde{\partial}_+ \omega_1 = -2\sin\frac{\omega_1}{2} = \tilde{\partial}_- \omega_1$$

or, $\omega_1 = \omega_1(\tilde{x}^+ + \tilde{x}^-)$ \qquad\qquad (8.32)

To determine the form of the solution explicitly, let us note that

$$\tilde{\partial}_+ \omega_1 = -2\sin\frac{\omega_1}{2} = -4\sin\frac{\omega_1}{4}\cos\frac{\omega_1}{4}$$

so that

$$\sec^2\frac{\omega_1}{4}\,\tilde{\partial}_+\omega_1 = -4\tan\frac{\omega_1}{4}$$

or, $\tilde{\partial}_+\left(\tan\frac{\omega_1}{4}\right) = -\tan\frac{\omega_1}{4}$ \qquad\qquad (8.33)

This equation can be readily integrated and recalling the form of $\omega_1(\tilde{x}^+, \tilde{x}^-)$ from Eq. (8.32), we see that

$$\tan\frac{\omega_1}{4} = C\,\exp(-(\tilde{x}^+ + \tilde{x}^-))$$

$$= C\,\exp\left(-(ax^+ + \frac{1}{a}\,x^-)\right)$$

$$= C \exp\left(-\left(t\left(a + \frac{1}{a}\right) + x\left(a - \frac{1}{a}\right)\right)\right)$$

$$= C \exp\left(-\frac{1}{a}\left(t(a^2+1) + x(a^2-1)\right)\right)$$

$$= C \exp\left(\frac{-2(t+vx)}{(1-v^2)^{1/2}}\right) \tag{8.34}$$

where

$$v = \frac{a^2-1}{a^2+1} \tag{8.35}$$

and C is a constant.   Therefore, we have

$$\omega_1 = 4\tan^{-1}\left\{C \exp\left(\frac{-2(t+vx)}{(1-v^2)^{1/2}}\right)\right\} \tag{8.36}$$

and we see that we have obtained a solution of the Sine-
Gordon equation starting from the vacuum solution, $\omega=0$.
Furthermore, let us note that, for $v > 0$, this is nothing
other than a topological kink solution moving to the left
which is more commonly known as an antikink solution. Kinks
are topological soliton solutions and we see that the
Bäcklund transformations give a way of generating these from
the more trivial vacuum solutions.   This process can be
further carried out to generate more complicated solutions.
Normally, constructing solutions through such a procedure
would appear to be formidable if not for the fact that the
Bäcklund transformations satisfy the theorem of
permutability which simplifies life considerably.

Theorem of Permutability:

According to the theorem of permutability, two successive Bäcklund transformations are commutative. More quantitatively, if two Bäcklund transformations with distinct parameters a_1 and a_2 map a solution ω_0 to another solution ω_{12}, then their order is irrelevant. Namely, if

$$\omega_0 \xrightarrow{\quad a_1 \quad} \omega_1 \xrightarrow{\quad a_2 \quad} \omega_{12} \tag{8.37}$$

and

$$\omega_0 \xrightarrow{\quad a_2 \quad} \omega_2 \xrightarrow{\quad a_1 \quad} \omega_{21} \tag{8.38}$$

then

$$\omega_{12} = \omega_{21} \tag{8.39}$$

If we now apply this theorem to the Sine-Gordon equation, we see from Eqs. (8.22) and (8.23) that

$$\partial_+(\omega_1 - \omega_0) = -2a_1 \sin\left(\frac{\omega_1 + \omega_0}{2}\right) \tag{8.40}$$

$$\partial_+(\omega_{12} - \omega_1) = -2a_2 \sin\left(\frac{\omega_{12} + \omega_1}{2}\right) \tag{8.41}$$

$$\partial_+(\omega_2 - \omega_0) = -2a_2 \sin\left(\frac{\omega_2 + \omega_0}{2}\right) \tag{8.42}$$

$$\partial_+(\omega_{12} - \omega_2) = -2a_1 \sin\left(\frac{\omega_{12} + \omega_2}{2}\right) \tag{8.43}$$

Adding Eqs. (8.40) and (8.41) as well as Eqs. (8.42) and (8.43) and subtracting one from the other, we obtain

$$2a_1\left(\sin\left(\frac{\omega_{12}+\omega_2}{2}\right) - \sin\left(\frac{\omega_1+\omega_0}{2}\right)\right) - 2a_2\left(\sin\left(\frac{\omega_{12}+\omega_1}{2}\right)\right.$$

$$\left. - \sin\left(\frac{\omega_2+\omega_0}{2}\right)\right) = 0$$

or, $$\cos\left(\frac{\omega_{12}+\omega_2+\omega_1+\omega_0}{4}\right)\left(a_1\sin\left(\frac{\omega_{12}-\omega_0+\omega_2-\omega_1}{4}\right)\right.$$

$$\left. - a_2\sin\left(\frac{\omega_{12}-\omega_0-\omega_2+\omega_1}{4}\right)\right) = 0$$

or, $$a_1\sin\left(\frac{\omega_{12}-\omega_0+\omega_2-\omega_1}{4}\right) - a_2\sin\left(\frac{\omega_{12}-\omega_0-\omega_2+\omega_1}{4}\right) = 0 \quad (8.44)$$

Using the properties of the trigonometric functions, this can be further simplified to

$$(a_1-a_2)\sin\left(\frac{\omega_{12}-\omega_0}{4}\right)\cos\left(\frac{\omega_2-\omega_1}{4}\right)$$

$$- (a_1+a_2)\cos\left(\frac{\omega_{12}-\omega_0}{4}\right)\sin\left(\frac{\omega_1-\omega_2}{4}\right) = 0$$

or, $$\tan\left(\frac{\omega_{12}-\omega_0}{4}\right) = \frac{a_1+a_2}{a_1-a_2}\tan\left(\frac{\omega_1-\omega_2}{4}\right) \quad (8.45)$$

In other words, the theorem of permutability allows one to construct a second order solution algebraically. Furthermore, this process can be carried out order by order

so that after construction of the first solution, we do not
have to go through the complicated formalism of quadratures.

The Bäcklund transformations are, therefore, of great
help in constructing solutions.   The difficulty lies, of
course, in first finding a Bäcklund transformation. There
exists a method, due to Clairin, to construct Bäcklund
transformations systematically.   However, it is not always
simple and straightforward.

Bäcklund Transformation for the KdV Equation:

We have already studied the Bäcklund transformations for
the KdV equation without quite recognizing them. These are
the Miura transformations or the Riccati relation. Let us
recall from Eqs. (6.4) and (6.6) that the generalized Miura
transformation

$$u + 6\lambda = v^2 + i\sqrt{6}\ v_x$$

leads to the generalized MKdV equation

$$v_t = (v^2 - 6\lambda)v_x + v_{xxx}$$

These can indeed be taken as the defining relations for the
Bäcklund transformations, if we rewrite them as

$$v_x = -\frac{i}{\sqrt{6}}\ (u + 6\lambda - v^2) \tag{8.46}$$

$$v_t = -\frac{i}{\sqrt{6}}\ u_{xx} - \frac{i}{3\sqrt{6}}\ u^2 + \frac{2i}{\sqrt{6}}\ \lambda u + \frac{24i}{\sqrt{6}}\ \lambda^2$$

$$+ \frac{1}{3}\ u_x v + \frac{i}{3\sqrt{6}}\ uv^2 - \frac{4i}{\sqrt{6}}\ \lambda v^2 \tag{8.47}$$

Differentiating Eq. (8.46) with respect to t and using the Eq. (8.47) to simplify, we obtain

$$v_{tx} = - \frac{i}{\sqrt{6}} u_t + \frac{2i}{\sqrt{6}} vv_t$$

$$= - \frac{i}{\sqrt{6}} u_t + \frac{1}{3} u_{xx}v + \frac{1}{9} u^2 v - \frac{2}{3} \lambda uv$$

$$- 8\lambda^2 v + \frac{2i}{3\sqrt{6}} u_x v^2 - \frac{1}{9} uv^3 + \frac{4}{3} \lambda v^3 \qquad (8.48)$$

Note that we are tacitly assuming that λ is independent of space and time. We, of course, know that it is true since λ is the eigenvalue of the Schrödinger equation. Another way to see this is to note that it is the parameter of a Galilean transformation.

We can also differentiate Eq. (8.47) with respect to x and use the Eq. (8.46) to simplify. In this way we get

$$v_{xt} = - \frac{i}{\sqrt{6}} u_{xxx} - \frac{2i}{3\sqrt{6}} uu_x + \frac{2i}{\sqrt{6}} \lambda u_x + \frac{1}{3} u_{xx}v$$

$$+ \frac{i}{3\sqrt{6}} u_x v^2 - \frac{i}{\sqrt{6}} \left(\frac{1}{3} u_x + \frac{2i}{3\sqrt{6}} uv - \frac{8i}{\sqrt{6}} \lambda v \right) (u + 6\lambda - v^2)$$

$$= - \frac{i}{\sqrt{6}} u_{xxx} - \frac{i}{\sqrt{6}} uu_x + \frac{1}{3} u_{xx}v + \frac{1}{9} u^2 v$$

$$- \frac{2}{3} \lambda uv - 8\lambda^2 v + \frac{2i}{3\sqrt{6}} u_x v^2 - \frac{1}{9} uv^3 + \frac{4}{3} \lambda v^3 \quad (8.49)$$

The integrability condition for the defining relations in Eqs. (8.46) and (8.47) follows from (8.48) and (8.49) and gives the KdV equation, namely,

$$u_t = uu_x + u_{xxx}$$

The evolution equation for v is one of the defining relations of the Bäcklund transformation and can be shown to be the MKdV equation upon eliminating u through Eq. (8.46).

This, therefore, is an example of the Bäcklund transformation which connects a solution of the KdV equation to that of the MKdV equation. We can also ask whether there exists a Bäcklund transformation which relates one solution of the KdV equation to another. If such a transformation exists, it will be potentially quite useful since we could then in principle construct multi-soliton solutions starting from a single soliton solution which we have already worked out. The answer to the above question, surprisingly, is yes and the Bäcklund transformation, in this case, can be constructed as follows.

Let us introduce a new variable $\omega(x,t)$, defined by

$$u(x,t) = \frac{\partial \omega(x,t)}{\partial x} \qquad (8.50)$$

The KdV equation in terms of the new variable becomes

$$u_t = uu_x + u_{xxx}$$

$$\text{or,} \quad \omega_t = \frac{1}{2} \omega_x^2 + \omega_{xxx} \tag{8.51}$$

Let us also note from Eq. (6.6) that if $v(x,t)$ is a solution of the MKdV equation for a given value of λ, then so is $-v(x,t)$. Each of these solutions, of course, leads to a unique solution of the KdV equation. Denoting by $u(x,t)$ and $\tilde{u}(x,t)$, the solutions obtained from $v(x,t)$ and $-v(x,t)$ respectively, the relations (8.46) and (8.47) become

$$v_x = - \frac{i}{\sqrt{6}} (u+6\lambda-v^2)$$

$$-v_x = - \frac{i}{\sqrt{6}} (\tilde{u}+6\lambda-v^2)$$

$$v_t = - \frac{i}{\sqrt{6}} u_{xx} - \frac{i}{3\sqrt{6}} u^2 + \frac{2i}{\sqrt{6}} \lambda u + \frac{24i}{\sqrt{6}} \lambda^2 \tag{8.52}$$

$$+ \frac{1}{3} u_x v + \frac{i}{3\sqrt{6}} uv^2 - \frac{4i}{\sqrt{6}} \lambda v^2$$

$$-v_t = - \frac{i}{\sqrt{6}} \tilde{u}_{xx} - \frac{i}{3\sqrt{6}} \tilde{u}^2 + \frac{2i}{\sqrt{6}} \lambda\tilde{u} + \frac{24i}{\sqrt{6}} \lambda^2$$

$$- \frac{1}{3} \tilde{u}_x v + \frac{i}{3\sqrt{6}} \tilde{u}v^2 - \frac{4i}{\sqrt{6}} \lambda v^2$$

From these relations we obtain

$$v_x = - \frac{i}{2\sqrt{6}} (u-\tilde{u}) \tag{8.53}$$

$$\frac{1}{2} (u+\tilde{u}) = v^2 - 6\lambda \tag{8.54}$$

$$v_t = -\frac{i}{2\sqrt{6}}(u-\tilde{u})_{xx} - \frac{i}{4\sqrt{6}}(u^2-\tilde{u}^2) \qquad (8.55)$$

Recalling now that $u = \omega_x$ and $\tilde{u} = \tilde{\omega}_x$, we see that Eq. (8.53) leads to the identification

$$v = -\frac{i}{2\sqrt{6}}(\omega-\tilde{\omega}) \qquad (8.56)$$

Using this then Eqs. (8.54) and (8.55) define a Bäcklund transformation between the variables ω and $\tilde{\omega}$ in the form

$$(\omega+\tilde{\omega})_x = -(12\lambda + \frac{1}{12}(\omega-\tilde{\omega})^2) \qquad (8.57)$$

$$(\omega-\tilde{\omega})_t = (\omega-\tilde{\omega})_{xxx} + \frac{1}{2}(\omega-\tilde{\omega})_x(\omega+\tilde{\omega})_x \qquad (8.58)$$

Let us note here some identities, following from Eq. (8.57), which would be quite useful in the analysis of the integrability of these relations.

$$(\omega+\tilde{\omega})_{xx} = -\frac{1}{6}(\omega-\tilde{\omega})(\omega-\tilde{\omega})_x$$

$$\qquad (8.59)$$

$$(\omega+\tilde{\omega})_{xxxx} = -\frac{1}{6}(\omega-\tilde{\omega})(\omega-\tilde{\omega})_{xxx} - \frac{1}{2}(\omega-\tilde{\omega})_x(\omega-\tilde{\omega})_{xx}$$

Differentiating Eq. (8.57) with respect to t, we obtain

$$(\omega+\tilde{\omega})_{tx} = -\frac{1}{6}(\omega-\tilde{\omega})(\omega-\tilde{\omega})_t$$

$$= -\frac{1}{6}(\omega-\tilde{\omega})(\omega-\tilde{\omega})_{xxx} - \frac{1}{12}(\omega-\tilde{\omega})(\omega-\tilde{\omega})_x(\omega+\tilde{\omega})_x \quad (8.60)$$

Using the identities from Eq. (8.59), this becomes

$$(\omega+\tilde{\omega})_{tx} = (\omega+\tilde{\omega})_{xxxx} + \omega_x\omega_{xx} + \tilde{\omega}_x\tilde{\omega}_{xx} \qquad (8.61)$$

Similarly, differentiating Eq. (8.58) with respect to x, we obtain

$$(\omega-\tilde{\omega})_{xt} = (\omega-\tilde{\omega})_{xxxx} + \frac{1}{2}(\omega-\tilde{\omega})_{xx}(\omega+\tilde{\omega})_x + \frac{1}{2}(\omega-\tilde{\omega})_x(\omega+\tilde{\omega})_{xx}$$

$$= (\omega-\tilde{\omega})_{xxxx} + \omega_x\omega_{xx} - \tilde{\omega}_x\tilde{\omega}_{xx} \qquad (8.62)$$

Thus the integrability conditions for the Bäcklund transformations defined in Eqs. (8.57) and (8.58) become

$$\omega_{xt} = \omega_{xxxx} + \omega_x\omega_{xx}$$

$$\text{or,} \quad \omega_t = \frac{1}{2}\omega_x^2 + \omega_{xxx}$$

and   (8.63)

$$\tilde{\omega}_{xt} = \tilde{\omega}_{xxxx} + \tilde{\omega}_x\tilde{\omega}_{xx}$$

$$\text{or,} \quad \tilde{\omega}_t = \frac{1}{2}\tilde{\omega}_x^2 + \tilde{\omega}_{xxx}$$

Thus both ω and $\tilde{\omega}$ satisfy the same equation which as we see from Eq. (8.51) is nothing other than the KdV equation in these variables.   This   set   of   Bäcklund transformations, therefore, relates one solution of the KdV equation to another and hence has the potential for constructing new solutions of the system.

Soliton Solutions:

Since we know that $\omega = 0$ is the trivial solution of the KdV equation, let us ask whether the Bäcklund transformations can lead to any interesting solution in such a case. For $\omega = 0$, the Bäcklund transformations of Eqs. (8.57) and (8.58) reduce to

$$\tilde{\omega}_x = -12\lambda - \frac{1}{12}\tilde{\omega}^2 \tag{8.64}$$

$$\tilde{\omega}_t = \tilde{\omega}_{xxx} + \frac{1}{2}\tilde{\omega}_x^2 \tag{8.65}$$

The second equation is identical to Eq. (8.63) and, in this case, is simply the statement that $\tilde{\omega}$ is a solution of the KdV equation. However, let us note from Eq. (8.64) that

$$\tilde{\omega}_{xxx} = -\frac{1}{6}\tilde{\omega}_x^2 + \frac{1}{36}\tilde{\omega}^2\tilde{\omega}_x \tag{8.66}$$

Hence the evolution equation, Eq. (8.65), in this case takes the form

$$\tilde{\omega}_t = \tilde{\omega}_{xxx} + \frac{1}{2}\tilde{\omega}_x^2 = -\frac{1}{6}\tilde{\omega}_x^2 + \frac{1}{36}\tilde{\omega}^2\tilde{\omega}_x + \frac{1}{2}\tilde{\omega}_x^2$$

$$= \frac{1}{3}\tilde{\omega}_x^2 + \frac{1}{36}\tilde{\omega}^2\tilde{\omega}_x = \frac{1}{3}\tilde{\omega}_x\left(\tilde{\omega}_x + \frac{1}{12}\tilde{\omega}^2\right)$$

which upon using Eq. (8.64) becomes

$$\tilde{\omega}_t = -4\lambda\tilde{\omega}_x \qquad (8.67)$$

Thus the solution obtained from the trivial vacuum solution through the Bäcklund transformation corresponds to a travelling wave travelling with the speed (compare with Eq. (7.34))

$$c = -4\lambda \qquad (8.68)$$

Furthermore, if we recall that the eigenvalue of the Schrödinger equation, namely λ, for the bound states takes negative values, then such a wave moves to the left. We readily recognize this as the single soliton solution. In fact, if we choose

$$\omega(x,t) = 12\sqrt{-\lambda} \tanh \sqrt{-\lambda} (x-4\lambda t) \qquad (8.69)$$

so that

$$u(x,t) = \omega_x(x,t) = -12\lambda \operatorname{sech}^2 \sqrt{-\lambda} (x-4\lambda t) \qquad (8.70)$$

then we see that Eqs. (8.64), (8.65) and (8.67) are completely satisfied. But this is precisely the soliton solution (see Eqs. (4.34) and (4.49)) we had constructed earlier. This is indeed encouraging since it suggests that the multi-soliton solutions can possibly be constructed by successive application of the Bäcklund transformations.

We have seen earlier in connection with the Sine-Gordon equation that the theorem of permutability plays an important role in constructing higher order solutions through Bäcklund transformations.   Let us examine what this theorem says in the present case.   Let us suppose

$$\omega \xrightarrow{\ \lambda_1\ } \omega_1 \xrightarrow{\ \lambda_2\ } \omega_{12}$$

$$\omega \xrightarrow{\ \lambda_2\ } \omega_2 \xrightarrow{\ \lambda_1\ } \omega_{21}$$

Then, of course, the theorem of permutability requires

$$\omega_{12} = \omega_{21}$$

Quantitatively, from the Bäcklund transformations, in Eqs. (8.57), then, we obtain

$$(\omega + \omega_1)_x = -12\lambda_1 - \frac{1}{12}\ (\omega - \omega_1)^2 \tag{8.71}$$

$$(\omega_1 + \omega_{12})_x = -12\lambda_2 - \frac{1}{12}\ (\omega_1 - \omega_{12})^2 \tag{8.72}$$

$$(\omega + \omega_2)_x = -12\lambda_2 - \frac{1}{12}\ (\omega - \omega_2)^2 \tag{8.73}$$

$$(\omega_2 + \omega_{12})_x = -12\lambda_1 - \frac{1}{12}\ (\omega_2 - \omega_{12})^2 \tag{8.74}$$

Subtracting Eq. (8.71) from Eq. (8.72) and Eq. (8.73) from Eq. (8.74) and then the result of one from the other we obtain

$$\frac{1}{12}\left[(\omega-\omega_1)^2-(\omega_1-\omega_{12})^2-(\omega-\omega_2)^2+(\omega_2-\omega_{12})^2\right] = 24(\lambda_2-\lambda_1)$$

or, $(\omega - \omega_{12})(\omega + \omega_{12} - 2\omega_1 - \omega - \omega_{12} + 2\omega_2) = 288(\lambda_2-\lambda_1)$

or, $\omega_{12} = \omega - \dfrac{144(\lambda_2-\lambda_1)}{(\omega_2-\omega_1)}$ (8.75)

Note that this expression is symmetric under interchange of the 1 and 2 indices as the permutability theorem would require and that

$$u_{12}(x,t) = \frac{\partial \omega_{12}(x,t)}{\partial x}$$ (8.76)

The theorem of permutability allows construction of the higher order solutions algebraically at every given order. Thus the general expression for the solution at the nth step can be written as

$$\omega_{(n)} = \omega_{(n-2)} - \frac{144(\lambda_n-\lambda_{n-1})}{\omega'_{(n-1)}-\omega_{(n-1)}} \qquad n > 1 \quad (8.77)$$

where

$$\omega_{(n)} = \omega_{\{k_1,k_2\ldots k_n\}} = \omega_{12\ldots n}$$

(8.78)

$$\omega'_{(n)} = \omega_{\{k_1\ldots k_{n-1},k_{n+1}\}} = \omega_{12\ldots n-1,n+1}$$

and

$$\omega_{(0)} = \omega \tag{8.79}$$

For example,

$$\omega_{123} = \omega_1 - \frac{144(\lambda_3 - \lambda_2)}{\omega_{13} - \omega_{12}} \tag{8.80}$$

and since we know the form of the 2nd step solutions from Eq. (8.75), namely,

$$\omega_{12} = \omega - \frac{144(\lambda_2 - \lambda_1)}{\omega_2 - \omega_1}$$

$$\tag{8.81}$$

$$\omega_{13} = \omega - \frac{144(\lambda_3 - \lambda_1)}{(\omega_3 - \omega_1)}$$

we can determine explicitly the form of ω_{123} in terms of the 1st step solutions to be

$$\omega_{123} = \frac{\lambda_1 \omega_1 (\omega_2 - \omega_3) + \lambda_2 \omega_2 (\omega_3 - \omega_1) + \lambda_3 \omega_3 (\omega_1 - \omega_2)}{\lambda_1 (\omega_2 - \omega_3) + \lambda_2 (\omega_3 - \omega_1) + \lambda_3 (\omega_1 - \omega_2)}$$

The permutation symmetry of this solution in the indices 1, 2 and 3 is manifest.

We have already seen that if

$$\omega_{(0)} = \omega = 0 \tag{8.82}$$

then at least the first order (step) solutions correspond to single soliton solutions. The regular solutions for the soliton equation, as we have seen in Eq. (8.69), take the form

$$\omega(x,t) = 12\sqrt{-\lambda} \ \tanh \ \sqrt{-\lambda} \ (x-4\lambda t)$$

But a singular solution also satisfies the same equation, namely,

$$\omega^*(x,t) = 12\sqrt{-\lambda} \ \coth \ \sqrt{-\lambda} \ (x-4\lambda t) \qquad (8.83)$$

also satisfies Eq. (8.64) in that

$$\omega_x^* = -12\lambda - \frac{1}{12} \ \omega^{*2} \qquad\qquad (8.84)$$

It turns out that to obtain regular solutions through the application of Bäcklund transformations, we do not always have to use regular solutions in the intermediate steps. This is easily seen in the 2nd step solution obtained from the vacuum, namely, if we choose

$$\omega_{12} = - \ \frac{144(\lambda_2-\lambda_1)}{\omega_2^*-\omega_1} \qquad \lambda_1 > \lambda_2 \qquad (8.85)$$

then the denominator never vanishes and hence the solution is regular. In fact, let us note that because the λ's are negative, our ordering says that while λ_1 can vanish, λ_2 will not. In the limit $\lambda_1 = 0$

$$\omega_{12}(\lambda_1 = 0) = -\frac{144\lambda_2}{\omega_2^*}$$

$$= 12\sqrt{-\lambda_2} \, \tanh \sqrt{-\lambda_2} \, (x - 4\lambda_2 t) = \omega_2 \quad (8.86)$$

This is, of course, regular.   It is also clear that had we used the regular solution ω_2 in the intermediate step, this limit would have led to a singular solution. Thus we see that to obtain regular solutions, not only do we have to use singular solutions in the intermediate steps, but that we must also order the parameters of the transformations in some specific way.

Next we will analyze the asymptotic behavior of the 2nd step solution obtained from the vacuum. And to simplify our task, we will write

$$\omega_1 = 12\sqrt{-\lambda_1} \, \tanh \xi_1$$

$$\omega_2 = 12\sqrt{-\lambda_2} \, \tanh \xi_2 \qquad (8.87)$$

where

$$\xi_1 = \sqrt{-\lambda_1} \, (x - 4\lambda_1 t)$$

$$\xi_2 = \sqrt{-\lambda_2} \, (x - 4\lambda_2 t) \qquad (8.88)$$

Let us further introduce the notation

$$\tanh \gamma = \frac{\sqrt{-\lambda_1}}{\sqrt{-\lambda_2}} \qquad (8.89)$$

From the form

$$\omega_{12} = -\frac{144(\lambda_2 - \lambda_1)}{\omega_2^* - \omega_1}$$

we see that as $\xi_2 \to \pm\infty$

$$\omega_{12} \longrightarrow \pm 12\sqrt{-\lambda_1}\,\coth\gamma + 12\sqrt{-\lambda_1}\,\tanh\,(\xi_1 \mp \gamma) \qquad (8.90)$$

Similarly, as $\xi_1 \to \pm\infty$

$$\omega_{12} \longrightarrow \pm 12\sqrt{-\lambda_2}\,\tanh\gamma + 12\sqrt{-\lambda_2}\,\tanh\,(\xi_2 \mp \gamma) \qquad (8.91)$$

Thus we see that

$$u_{12} = \omega_{12,x} \xrightarrow[|\xi_2| \to \infty]{} - 12\lambda_1 \mathrm{sech}^2(\xi_1 \mp \gamma)$$

$$\qquad\qquad (8.92)$$

$$\xrightarrow[|\xi_1| \to \infty]{} - 12\lambda_2 \mathrm{sech}^2(\xi_2 \mp \gamma)$$

This shows that u_{12} is indeed a pure two soliton solution with asymptotic phases $\dfrac{\gamma}{\sqrt{-\lambda_1}}$ and $\dfrac{\gamma}{\sqrt{-\lambda_2}}$ respectively. This result is quite significant since it tells us that starting with the vacuum solution, we can construct the multi-soliton solutions through repeated application of the Bäcklund transformations, increasing the soliton number by unity at each successive step. That this is true can actually be checked recursively.

References:

Bäcklund, A. V., Lund Univ. Arsskrift $\underline{10}$ (1875).

Bäcklund, A. V., Math. Ann. $\underline{9}$, 297 (1876).

Chen, H-H in Bäcklund Transformations, Ed. A. Dold and B. Eckmann, Springer-Verlag, 1974.

Clairin, J., Ann. Sci. Ecole Norm. Sup. 3^e Ser. Suppl. $\underline{19}$ (1902).

Dold, A. and B. Eckmann, Bäcklund Transformations, Springer-Verlag, 1974.

Eisenhart, L. P., A Treatise on the Differential Geometry of Curves and Surfaces, Dover, 1960.

Forsyth, A. R., Theory of Differential Equations, Vol. VI, Dover, 1959.

Liouville, J., J. Math. Pures Appl. $\underline{18}$, 71 (1853).

Newell, A. C. in Bäcklund Transformations, Ed. A. Dold and B. Eckmann, Springer-Verlag, 1974.

Wadati, M., J. Phys. Sco. Japan $\underline{36}$, 1498 (1974).

Wadati, M., H. Sanuki and K. Konno, Prog. Theor. Phys. $\underline{53}$, 419 (1975).

Wahlquist, H. D. and F. B. Estabrook, Phys. Rev. Lett. $\underline{31}$, 1386 (1973).

CHAPTER 9

GEOMETRICAL APPROACH TO INTEGRABLE MODELS

So far we have studied many distinctly fascinating aspects of integrable models. However, most of these concepts appear to be quite disjoint. For example, it is not yet clear whether the existence of a dual Poisson bracket structure in such models is in any way responsible for the existence of the Lax operators. Secondly, we do not understand whether there exists any direct relation between the conserved quantities and the Lax operators or why functional recursion relations exist between the conserved quantities. A geometrical approach to the study of the integrable models unifies some of these concepts and hence clarifies some of our questions. In this section, we would undertake such a study. However, for simplicity we would restrict ourselves to the study of a finite dimensional system. This also allows us to introduce a finite dimensional integrable system, namely, the Toda lattice.

Symplectic Geometry:

Let us recapitulate the basic elements of the phase space geometry which is meaningful for a Hamiltonian system. We have said earlier in Chapter 1 that the geometry of the phase space is a symplectic geometry.

Formally, a symplectic manifold M is defined to be a manifold with a preferred two form which is nondegenerate and is closed. The properties of the manifold follow from this definition.

For example, let us recall that a two form, $f \in M$, is said to be nondegenerate if for every vector field $V \in M$ and a fixed vector field $W \in M$,

$$f(V,W) = 0 \tag{9.1}$$

implies that

$$W = 0 \tag{9.2}$$

Let us assume that the finite dimensional manifold M is spanned by general coordinates y^μ.   In this basis then, we can write

$$V = v^\mu \frac{\partial}{\partial y^\mu} = v^\mu \, \partial_\mu$$

$$W = w^\mu \cdot \frac{\partial}{\partial y^\mu} = w^\mu \, \partial_\mu \tag{9.3}$$

$$f = \frac{1}{2} \, f_{\mu\nu}(y) dy^\mu \wedge dy^\nu$$

Furthermore, using the normalization

$$\left(dy^\mu, \, \partial_\nu\right) = \delta^\mu_\nu \tag{9.4}$$

the contraction of the two form f with the vector fields V and W is seen to give

$$f(V,W) = V^{\mu} f_{\mu\nu} W^{\nu\cdot} \qquad (9.5)$$

The conditions for nondegeneracy, namely Eqs. (9.1) and (9.2), then simply imply that the antisymmetric coefficient matrix $f_{\mu\nu}(y)$ is nonsingular. For this to be true, the coefficient matrix and hence the manifold M must be even dimensional. Thus a symplectic manifold is by definition even dimensional.

Furthermore, since $f_{\mu\nu}(y)$ is nonsingular, let $f^{\mu\nu}(y)$ be its inverse satisfying

$$f_{\mu\lambda}(y) \ f^{\lambda\nu}(y) = \delta^{\nu}_{\mu} = f^{\nu\lambda}(y)f_{\lambda\mu}(y) \qquad (9.6)$$

The inverse, $f^{\mu\nu}(y)$, allows us to define a bracket between two scalar functions on the manifold as

$$\{p(y),q(y)\} = f^{\mu\nu}(y)\partial_{\mu}p(y)\partial_{\nu}q(y) \qquad (9.7)$$

Here we have used $\partial_{\mu} = \dfrac{\partial}{\partial y^{\mu}}$ and the summation convention over repeated indices. This bracket is by definition antisymmetric since $f^{\mu\nu}$ is, namely,

$$\{p(y),q(y)\} = f^{\mu\nu}\partial_{\mu}p\partial_{\nu}q$$

$$= -f^{\nu\mu}\partial_{\nu}q\partial_{\mu}p = -\{q(y),p(y)\} \qquad (9.8)$$

$f^{\mu\nu}(y)$ also allows us to construct a very special class of

vector fields, known as the Hamiltonian vector fields, from
given scalar functions as

$$X_p = f^{\mu\nu}\partial_\mu p \, \partial_\nu \tag{9.9}$$

so that in terms of these vector fields, we can write

$$\{p(y), q(y)\} = -f(X_p, X_q) \tag{9.10}$$

where the right hand side denotes the natural product of the
two form f with the two Hamiltonian vector fields X_p and X_q.
We can also calculate the contraction of the three form df
with three arbitrary Hamiltonian vector fields X_p, X_q and X_r
in the following way.

$$df(X_p, X_q, X_r) = \partial_\lambda f(X_p, X_q)(dy^\lambda, X_r) + \partial_\lambda f(X_q, X_r)(dy^\lambda, X_p)$$

$$+ \partial_\lambda f(X_r, X_p)(dy^\lambda, X_q)$$

$$= f^{\mu\nu}\big(\partial_\nu f(X_p, X_q)\partial_\mu r + \partial_\nu f(X_q, X_r)\partial_\mu p$$

$$+ \partial_\nu f(X_r, X_p)\partial_\mu q\big)$$

Using Eq. (9.10) then, we obtain

$$df(X_p, X_q, X_r) = -\{p, \{q, r\}\} - \{q, \{r, p\}\} - \{r, \{p, q\}\} \tag{9.11}$$

Thus if the two form f is closed, namely, if

$$df = 0 \tag{9.12}$$

then the Jacobi identity associated with the bracket is satisfied and, consequently, the bracket in Eq. (9.7) can be identified with the Poisson bracket of the functions p(y) and q(y). A symplectic manifold, therefore, has a natural Poisson bracket structure defined in terms of the inverse of the symplectic structure $f_{\mu\nu}(y)$.

Note that the tensors $f_{\mu\nu}(y)$ and $f^{\mu\nu}(y)$ can be thought of as the components of the covariant and contravariant metric tensors of the symplectic manifold. This, of course, follows from the fact that a two form maps two vector fields into the reals. That is,

$$f: \quad V \times V \longrightarrow R \tag{9.13}$$

That $f_{\mu\nu}(y)$ and $f^{\mu\nu}(y)$ indeed transform like the components of a second rank tensor can also be directly verified as follows. Let us note from Eq. (9.7) that

$$\{y^{\mu}, y^{\nu}\} = f^{\mu\nu}(y) \tag{9.14}$$

Under a diffeomorphism

$$y^{\mu} \rightarrow \bar{y}^{\mu}(y)$$

$$f^{\mu\nu}(y) \rightarrow \bar{f}^{\mu\nu}(\bar{y}) \tag{9.15}$$

such that

$$\{\bar{y}^{\mu}, \bar{y}^{\nu}\} = \bar{f}^{\mu\nu}(\bar{y}) \tag{9.16}$$

Using (9.7), on the other hand, we obtain

$$\bar{f}^{\mu\nu}(\bar{y}) = \{\bar{y}^{\mu}, \bar{y}^{\nu}\} = f^{\lambda\rho}(y)\partial_{\lambda}\bar{y}^{\mu}\partial_{\rho}\bar{y}^{\nu}$$

$$\text{or,} \quad \bar{f}^{\mu\nu}(\bar{y}) = f^{\lambda\rho}(y) \frac{\partial\bar{y}^{\mu}}{\partial y^{\lambda}} \frac{\partial\bar{y}^{\nu}}{\partial y^{\rho}} \tag{9.17}$$

This is indeed the transformation law for the components of a second rank contravariant tensor. The transformation law for $f_{\mu\nu}(y)$ follows since it is the inverse of $f^{\mu\nu}(y)$.

Let us note here that $f_{\mu\nu}(y)$ and $f^{\mu\nu}(y)$ are known as the covariant and contravariant components of the symplectic metric of the manifold. Note that such a metric tensor is antisymmetric as opposed to the metric tensor of a Riemannian manifold which is symmetric. The symplectic metric cannot, therefore, be used to define lengths. But $f^{\mu\nu}(y)$ and $f_{\mu\nu}(y)$ can still be used to raise or lower indices. Thus for example, the contragradient of a scalar function is given by

$$\partial^{\mu}p(y) = f^{\mu\nu}(y)\partial_{\nu}p(y) \tag{9.18}$$

which we recognize as the components of the Hamiltonian field X_p defined in Eq. (9.9) up to a sign. In fact, note that the above relation implies that

$$dp = \partial_{\mu}p \, dy^{\mu} = f_{\mu\nu}(y)\partial^{\nu}p \, dy^{\mu} = i_{X_p} f = f(X_p, \cdot)$$

where i_{X_p} is the standard contraction operation. This indeed brings out the very special nature of the Hamiltonian vector fields, namely, they generate isometries of a symplectic manifold. This can be checked by recalling that the Lie derivative L_X with respect to a vector field X can be written as

$$L_X = i_X \cdot d + d \cdot i_X \qquad (9.20)$$

from which it follows that

$$L_{X_p} f = i_{X_p} \cdot df + d \cdot i_{X_p} f = d(dp) = 0 \qquad (9.21)$$

Here the first term vanishes by Eq. (9.12) because the two form f is closed. Thus we see that the flow generated by the Hamiltonian vector fields leaves the symplectic metric form invariant. Conversely, a transformation which leaves the symplectic metric form invariant is called a symplectic diffeomorphism and is generated by a Hamiltonian vector field. (These are generalizations of the canonical transformations which leave the canonical Poisson bracket relations invariant.) Note that for a Hamiltonian vector field X_p, the flow is given by the Hamilton's equation with p(y) playing the role of the Hamiltonian, namely,

$$\dot{y}^\mu = \{y^\mu, p(y)\} = f^{\mu\nu} \partial_\nu p(y) \qquad (9.22)$$

Let us note here that $f_{\mu\nu}(y)$ and $f^{\mu\nu}(y)$ are in general functions of the coordinates y^μ.   However, if for a system there exist global coordinates   such that $f_{\mu\nu}(y)$ and $f^{\mu\nu}(y)$ take the form (in terms of N×N blocks when $\mu,\nu = 1,2\ldots2N$)

$$f_{\mu\nu} = \begin{pmatrix} 0 & -I \\ I & 0 \end{pmatrix} \quad , \quad f^{\mu\nu} = \begin{pmatrix} 0 & I \\ -I & 0 \end{pmatrix} \qquad (9.23)$$

then such coordinates   are called canonical coordinates and the phase space, the canonical phase space. In general, however, if the system cannot be described by global canonical coordinates, then $f_{\mu\nu}$ and $f^{\mu\nu}$ would be coordinate dependent. Darboux's theorem, in such a case, guarantees that we can at least choose a local coordinate basis in which the symplectic metric takes the canonical form of Eq. (9.23).

Integrabel Models:

    The properties described above hold for any symplectic manifold. The phase space of an integrable model, on the other hand, must correspond to a very special symplectic manifold since among other things it possesses at least a dual Poisson bracket structure.   To study the properties of such a manifold, let us assume that it possesses at least two distinct two forms which are both nondegenerate and closed. Each two form would give rise to its own Poisson bracket thereby leading to the dual Poisson bracket structure.   Let us   note that while the differential geometric language allows a compact description, in what follows, we would work out formulae in tensorial components for clarity.

One way of expressing the existence of two distinct symplectic structures is to require that the same dynamical equation be described by two distinct first order Lagrangians. For example, let us assume that

$$L_0 = \theta_\mu^{(0)}(y)\dot{y}^\mu - H_0(y) \tag{9.24}$$

and

$$L = \theta_\mu(y)\dot{y}^\mu - H(y) \tag{9.25}$$

where $\mu = 1,2,\ldots.2N$ and $\dot{y}^\mu = \dfrac{dy^\mu}{dt}$, describe the same Hamiltonian equations. The Euler-Lagrange equations following from Eqs. (9.24) and (9.25) are

$$f_{\mu\nu}(y)\dot{y}^\nu = \partial_\mu H_0(y)$$

$$F_{\mu\nu}(y)\dot{y}^\nu = \partial_\mu H(y) \tag{9.26}$$

where

$$f_{\mu\nu}(y) = \partial_\mu\theta_\nu^{(0)}(y) - \partial_\nu\theta_\mu^{(0)}(y)$$

$$F_{\mu\nu}(y) = \partial_\mu\theta_\nu(y) - \partial_\nu\theta_\mu(y) \tag{9.27}$$

Clearly, then, the two forms

$$f = \frac{1}{2} f_{\mu\nu}(y) dy^{\mu} \wedge dy^{\nu}$$

$$(9.28)$$

$$F = \frac{1}{2} F_{\mu\nu}(y) dy^{\mu} \wedge dy^{\nu}$$

are closed since $f_{\mu\nu}$ and $F_{\mu\nu}$ automatically satisfy the Bianchi identity

$$\partial_{\lambda} f_{\mu\nu} + \partial_{\mu} f_{\nu\lambda} + \partial_{\nu} f_{\lambda\mu} = 0$$

$$(9.29)$$

$$\partial_{\lambda} F_{\mu\nu} + \partial_{\mu} F_{\nu\lambda} + \partial_{\nu} F_{\lambda\mu} = 0$$

Furthermore, they must also be nondegenerate since the two equations in Eq. (9.26) describe the same dynamical system. Consequently, $f_{\mu\nu}(y)$ and $F_{\mu\nu}(y)$ are the two distinct symplectic structures associated with the dynamical system. Let their inverses be $f^{\mu\nu}(y)$ and $F^{\mu\nu}(y)$ respectively so that

$$f_{\mu\nu}(y) f^{\nu\lambda}(y) = F_{\mu\nu}(y) F^{\nu\lambda}(y) = \delta_{\mu}^{\lambda} \qquad (9.30)$$

Each of the contravariant tensors gives rise to a definition of the Poisson bracket structure as

$$\{p(y), q(y)\}_{0} = f^{\mu\nu}(y) \partial_{\mu} p(y) \partial_{\nu} q(y)$$

$$(9.31)$$

$$\{p(y), q(y)\} = F^{\mu\nu}(y) \partial_{\mu} p(y) \partial_{\nu} q(y)$$

so that the two equations in Eq. (9.26) take the Hamiltonian form

$$\dot{y}^\mu = \{y^\mu, H_0\}_0 = f^{\mu\nu}(y)\partial_\nu H_0(y)$$

$$(9.32)$$

$$\dot{y}^\mu = \{y^\mu, H\} = F^{\mu\nu}(y)\partial_\nu H(y)$$

Furthermore, since the two equations of (9.32) describe the same dynamical system, we have

$$f^{\mu\nu}(y)\partial_\nu H_0(y) = F^{\mu\nu}(y)\partial_\nu H(y) \qquad (9.33)$$

This is, of course, the prototype of the recursion relation that we have seen among the conserved quantities of the KdV system in Eq. (3.58).

Given two distinct symplectic metric structures on the manifold we can construct a nontrivial (1,1) tensor naturally as

$$S_\mu{}^\nu(y) = F_{\mu\lambda}(y)f^{\lambda\nu}(y) \qquad (9.34)$$

(The inverse $f_{\mu\lambda}F^{\lambda\nu}$ is also a natural (1,1) tensor but considering one is sufficient for our purposes.) It is clear that although Darboux's theorem allows one of the symplectic metric structures to take the canonical form locally, $S_\mu{}^\nu(y)$ would always have a nontrivial coordinate dependence.

Let us also note from Eq. (9.32) that since

$$\partial_\mu H_0(y) = f_{\mu\nu}\dot{y}^\nu \qquad (9.35)$$

consistency of this equation would require

$$0 = \partial_\mu \partial_\nu H_0(y) - \partial_\nu \partial_\mu H_0(y)$$

$$= \partial_\mu(f_{\nu\lambda}\dot{y}^\lambda) - \partial_\nu(f_{\mu\lambda}\dot{y}^\lambda)$$

$$= (\partial_\mu f_{\nu\lambda} + \partial_\nu f_{\lambda\mu})\dot{y}^\lambda + f_{\nu\lambda}\partial_\mu\dot{y}^\lambda - f_{\mu\lambda}\partial_\nu\dot{y}^\lambda$$

$$= -\partial_\lambda f_{\mu\nu}\dot{y}^\lambda - \partial_\mu\dot{y}^\lambda f_{\lambda\nu} + \partial_\nu\dot{y}^\lambda f_{\lambda\mu}$$

$$= -\frac{df_{\mu\nu}(y)}{dt} - U_\mu{}^\lambda(y)f_{\lambda\nu}(y) + U_\nu{}^\lambda(y)f_{\lambda\mu}(y)$$

or,
$$\frac{df_{\mu\nu}(y)}{dt} = -U_\mu{}^\lambda(y)f_{\lambda\nu}(y) + U_\nu{}^\lambda(y)f_{\lambda\mu}(y) \qquad (9.36)$$

Here we have used the fact that the $f_{\mu\nu}$ tensor satisfies the Bianchi identity given in Eq. (9.29) and we have defined a new quantity

$$U_\mu{}^\nu(y) = \partial_\mu\dot{y}^\nu = \partial_\mu\left(f^{\nu\lambda}\partial_\lambda H_0(y)\right) = \partial_\mu\left(F^{\nu\lambda}\partial_\lambda H(y)\right) \qquad (9.37)$$

Similarly, from Eq. (9.32), requiring the consistency relation

$$\partial_\mu\partial_\nu H(y) - \partial_\nu\partial_\mu H(y) = 0 \qquad (9.38)$$

would lead to an equation

$$\frac{dF_{\mu\nu}(y)}{dt} = -U_\mu{}^\lambda F_{\lambda\nu} + U_\nu{}^\lambda F_{\lambda\mu} \tag{9.39}$$

involving the same $U_\mu{}^\nu$ tensor.

It is worth pointing out here that $f_{\mu\nu}(y)$ and $F_{\mu\nu}(y)$ are two distinct symplectic structures on the manifold and that the two Hamiltonians $H_0(y)$ and $H(y)$ generate their own flows. As we have seen before, Hamiltonian flows leave the symplectic structures form invariant and Eqs. (9.36) and (9.39) are merely statements of this fact.

The corresponding equations for the inverses $f^{\mu\nu}(y)$ and $F^{\mu\nu}(y)$ follow from (9.36) and (9.39) and have the form

$$\frac{df^{\mu\nu}(y)}{dt} = f^{\mu\lambda} U_\lambda{}^\nu - f^{\nu\lambda} U_\lambda{}^\mu$$

$$\tag{9.40}$$

$$\frac{dF^{\mu\nu}(y)}{dt} = F^{\mu\lambda} U_\lambda{}^\nu - F^{\nu\lambda} U_\lambda{}^\mu$$

But more interesting is the equation for the (1,1) tensor $S_\mu{}^\nu(y)$. Let us note that since $S_\mu{}^\nu(y)$ is constructed from tensors which remain form invariant under a Hamiltonian flow, it must also remain form invariant. Indeed, from its definition, and the equations (9.36), (9.39) and (9.40), we obtain

$$\frac{dS_\mu{}^\nu(y)}{dt} = S_\mu{}^\lambda(y) U_\lambda{}^\nu(y) - U_\mu{}^\lambda(y) S_\lambda{}^\nu(y) \tag{9.41}$$

We can also write it in an obvious matrix notation as

$$\frac{dS(y)}{dt} = [S(y), U(y)] \qquad (9.42)$$

We recognize Eq. (9.42) as a Lax equation (compare this with Eq. (6.23)) and in particular if $S(y)$ is linear in y, then clearly Eq. (9.42) gives a Lax representation of the original dynamical equations. Thus we see that the geometrical approach has given a systematic way of constructing the Lax operators starting from the dual Poisson bracket structure and the equation of motion. Furthermore, we also understand, geometrically, that the Lax equation is merely a statement of the form invariance of the natural (1,1) tensor on such a manifold under Hamiltonian flows.

The consequences following from the Lax equation in Eq. (9.42) are indeed profound. For example, let us define a set of quantities

$$K_n = \frac{1}{n} \text{Tr} S^n \qquad n = \pm 1, \pm 2 \ldots .$$

$$\qquad (9.43)$$

$$K_0 = \log|\det S|$$

Then, from the identity

$$\text{Tr}\left(P(S)\, \frac{dS}{dt}\right) = \text{Tr}\left(P(S)[S,U]\right) = 0 \qquad (9.44)$$

it follows that

$$\frac{dK_n}{dt} = 0 \qquad\qquad \forall\ n \qquad\qquad (9.45)$$

Therefore, all the quantities, K_n , constructed above would be automatically conserved under the evolution of the dynamical system.

However, not all such quantities can be functionally independent. In fact, let us note that since S is a 2N×2N matrix, it follows that all K_n with $n \geq 2N+1$ as well as all $(detS)^n K_{-n}$ for $n \geq 1$ can be expressed as polynomials of $K_1, K_2 \ldots K_{2N}$. Furthermore, not all of $K_1, K_2 \ldots K_{2N}$ can also be functionally independent since that would overconstrain the phase space and lead to no dynamics of the system. For integrability, we need only N functionally independent conserved quantities. Let us, therefore, assume that there are exactly N such quantities in the above set. But integrability further requires that these conserved quantities must also be in involution. It is not clear that this can be proved with the amount of information we have.

To prove involution, we look for additional structures on the symplectic manifold. Let us note that given a (1,1) tensor on the manifold, one can construct a natural (1,2) tensor as follows.

$$N^{\mu}_{\alpha\beta} = -N^{\mu}_{\beta\alpha}$$

$$= s_{\alpha}{}^{\lambda}\partial_{\lambda}s_{\beta}{}^{\mu} - s_{\beta}{}^{\lambda}\partial_{\lambda}s_{\alpha}{}^{\mu} - s_{\lambda}{}^{\mu}\left(\partial_{\alpha}s_{\beta}{}^{\lambda} - \partial_{\beta}s_{\alpha}{}^{\lambda}\right) \qquad (9.46)$$

That this is a tensor can be readily checked by replacing the ordinary derivatives with covariant derivatives with an arbitrary connection and showing that the connection terms indeed vanish. This tensor is known as the Nijenhuis torsion tensor and is widely studied in connection with complex manifolds. Let us simply note that it is quite analogous to the spin torsion tensor in an ordinary Riemannian manifold.

Since we know the time evolution of $S_\mu{}^\nu$, we can use Eq. (9.42) to derive the time evolution of the Nijenhuis torsion tensor of the manifold and it takes the form

$$\frac{dN^\mu_{\alpha\beta}}{dt} = -U_\alpha{}^\lambda N^\mu_{\lambda\beta} + U_\beta{}^\lambda N^\mu_{\lambda\alpha} + U_\lambda{}^\mu N^\lambda_{\alpha\beta} \qquad (9.47)$$

This is again a statement of the fact that the Nijenhuis torsion tensor is form invariant under a Hamiltonian flow and, consequently, a vanishing torsion stays zero for all times.

This is quite interesting, for if we look at the definition, Eq. (9.46), of the Nijenhuis tensor, we see that

$$N^\mu_{\alpha\beta}(S^{n-1})_\mu{}^\beta = S_\alpha{}^\lambda\left(\partial_\lambda S_\beta{}^\mu\right)(S^{n-1})_\mu{}^\beta - (S^n)_\mu{}^\lambda\partial_\lambda S_\alpha{}^\mu$$

$$- (S^n)_\lambda{}^\beta\left(\partial_\alpha S_\beta{}^\lambda - \partial_\beta S_\alpha{}^\lambda\right)$$

$$= S_\alpha{}^\lambda\left(\partial_\lambda S_\beta{}^\mu\right)(S^{n-1})_\mu{}^\beta - (S^n)_\lambda{}^\beta\partial_\alpha S_\beta{}^\lambda$$

$$= \frac{1}{n} S_\alpha{}^\lambda\partial_\lambda \mathrm{Tr} S^n - \frac{1}{n+1}\partial_\alpha \mathrm{Tr} S^{n+1}$$

$$\text{or,} \quad N^\mu_{\alpha\beta}(S^{n-1})_\mu{}^\beta = S_\alpha{}^\lambda\partial_\lambda K_n - \partial_\alpha K_{n+1} \qquad \forall \ n \quad (9.48)$$

Consequently, if the Nijenhuis torsion tensor vanishes we
see that

$$S_\alpha{}^\lambda \partial_\lambda K_n = \partial_\alpha K_{n+1}$$

$$\text{or,} \quad f^{\mu\nu} \partial_\nu K_n = F^{\mu\nu} \partial_\nu K_{n+1} \tag{9.49}$$

Therefore, we see that the vanishing Nijenhuis torsion
tensor leads to a recursion relation between the conserved
quantities, reminiscent of the recursion relation obtained
in Eq. (9.33) The proof of involution is quite simple now.
Namely,

$$\{K_m, K_n\}_0 = f^{\mu\nu} \partial_\mu K_m \partial_\nu K_n$$

$$= F^{\mu\nu} \partial_\mu K_m \partial_\nu K_{n+1}$$

$$= f^{\mu\nu} \partial_\mu K_{m-1} \partial_\nu K_{n+1}$$

$$\text{or,} \quad \{K_m, K_n\}_0 = \{K_{m-1}, K_{n+1}\}_0 \tag{9.50}$$

Assuming $m > n$, if we iterate this (m-n) times, this leads
to

$$\{K_m, K_n\}_0 = \{K_n, K_m\}_0 = 0 \qquad \forall \quad m, n \tag{9.51}$$

Similarly we can also show that the conserved quantities are
in involution with respect to the second Poisson bracket

structure, namely,

$$\{K_m, K_n\} = 0 \qquad\qquad \forall m, n \qquad (9.52)$$

This, therefore, shows that the vanishing of the Nijenhuis torsion tensor ensures integrability. Let us emphasize here that the vanishing of the Nijenhuis torsion tensor also plays a significant role in the study of the complex manifolds. But our $S_\mu{}^\nu(y)$ is not an almost complex structure because

$$S^2 \neq -1 \qquad\qquad\qquad\qquad (9.53)$$

In fact, such a condition would render our analysis uninteresting.

Let us note that since all of the K_n's are conserved and are in involution, they can be thought of as generating their own Hamiltonian flows given by

$$\frac{dy^\mu}{dt_n} = \{y^\mu, K_n\}_0 = \{y^\mu, K_{n+1}\} \qquad (9.54)$$

Equivalently,

$$\frac{dy^\mu}{dt_n} = f^{\mu\nu}\partial_\nu K_n = F^{\mu\nu}\partial_\nu K_{n+1} \qquad (9.55)$$

These flows would commute among themselves and would correspond to the higher order equations of the hierarchy.

Furthermore, we can also find the Lax equation for the higher order equations as well and they have the form

$$\frac{dS}{dt_n} = [S, U_n] \qquad (9.56)$$

where

$$(U_n)_\mu{}^\nu = \partial_\mu\left(f^{\nu\lambda}\partial_\lambda K_n\right) = \partial_\mu\left(F^{\nu\lambda}\partial_\lambda K_{n+1}\right) \qquad (9.57)$$

Thus we understand the higher order Lax equations as representing the form invariance of $S_\mu{}^\nu(y)$ under the higher order Hamiltonian flows. Furthermore, all of these higher order equations share the same conserved quantities and hence are also integrable.

In this section, we have tried to unify some of the seemingly different aspects of integrable models from a geometrical approach and have also tried to clarify some of the questions raised in the beginning of this chapter. Before applying these results to a systematic study of a new system, namely, the Toda lattice, let us see how all of this works for the familiar KdV example.

Example: Geometrical Approach to KdV

Here we will only derive the Lax equation following our discussion above. Let us note that the KdV is a continuum model and hence our formulae derived earlier need to be generalized with care. Formally, the two Poisson structures of the KdV equation are given by

$$F^{\mu\nu} \rightarrow D$$

$$\tag{9.58}$$

$$f^{\mu\nu} \rightarrow M = D^3 + \frac{1}{3} \, (Du + uD)$$

Their structures can be understood by going to the coordinate basis where

$$F(x,y) = \langle y|D|x \rangle = \frac{\partial}{\partial x} \, \delta(x-y)$$

$$\tag{9.59}$$

$$f(x,y) = \langle y|M|x \rangle$$

$$= \Big(\frac{\partial^3}{\partial x^3} + \frac{1}{3} \, (\frac{\partial}{\partial x} \, u(x) + u(x) \, \frac{\partial}{\partial x})\Big)\delta(x-y)$$

It is clear now that with appropriate boundary conditions, we can write

$$F^{-1}(x,y) = \langle y|D^{-1}|x \rangle = \epsilon(x-y) \tag{9.60}$$

where the alternating step function is defined as in Eq. (1.59) to be

$$\epsilon(x-y) = (\theta(x-y) - \frac{1}{2}) = -\epsilon(y-x) \tag{9.61}$$

so that

$$\frac{\partial \epsilon(x-y)}{\partial x} = \delta(x-y) = - \frac{\partial \epsilon(x-y)}{\partial y} \tag{9.62}$$

and

$$\int_{-\infty}^{\infty} dz F(x,z) F^{-1}(z,y) = \int_{-\infty}^{\infty} dz F^{-1}(x,z) F(z,y) \qquad (9.63)$$

$$= \delta(x-y)$$

$f^{-1}(x,y)$, on the other hand, cannot be expressed in a closed form.

It is obvious that in the present case the Poisson structures are operators and hence "S_{μ}^{ν}" needs to be defined carefully. A simple and careful choice turns out to be

$$S(x,y) = \int_{-\infty}^{\infty} dz F^{-1}(x,z) f(z,y)$$

which in the operator language takes the form

$$S = MD^{-1} = \left(D^3 + \frac{1}{3}(Du+uD)\right)D^{-1}$$

$$= D^2 + \frac{2}{3} u + \frac{1}{3}(Du)D^{-1} \qquad (9.64)$$

This is the Lax operator following from our earlier discussion of Eq. (9.42). We note that it is linear in the KdV variable u, but is nonlocal because of the D^{-1} term. If we restrict it to be local, then we obtain the more familiar Lax operator of Eq. (6.27) but with a different coefficient of the potential term. Let us, however, continue without any such restriction.

The KdV equation can be formally written as (see Eq. (7.1))

$$\dot{u} = K(u) = (D^3 u) + u(Du) \qquad (9.65)$$

so that the second Lax operator, following our earlier discussion in Eq. (9.37), is obtained to be

$$U = \frac{\delta \dot{u}}{\delta u} = \frac{\delta K(u)}{\delta u} = D^3 + Du \qquad (9.66)$$

Now the Lax equation

$$\frac{dS(x,y)}{dt} = \int_{-\infty}^{\infty} dz \Big(S(x,z)U(z,y) - U(x,z)S(z,y) \Big) \qquad (9.67)$$

takes the operator form

$$\frac{dS}{dt} = [U,S] \qquad (9.68)$$

We can calculate

$$SU = \Big(D^2 + \frac{2}{3} u + \frac{1}{3} (Du)D^{-1} \Big)(D^3 + Du)$$

$$= D^5 + D^3 u + \frac{2}{3} uD^3 + \frac{2}{3} uDu$$

$$+ \frac{1}{3} (Du)D^2 + \frac{1}{3} u(Du)$$

$$(9.69)$$

$$US = (D^3 + Du)\left(D^2 + \frac{2}{3} u + \frac{1}{3} (Du)D^{-1}\right)$$

$$= D^5 + \frac{2}{3} D^3 u + \frac{1}{3} D^3(Du)D^{-1} + DuD^2 + \frac{2}{3} Du^2$$

$$+ \frac{1}{3} Du(Du)D^{-1}$$

The Lax equation, therefore, upon simplification gives

$$\frac{dS}{dt} = [U,S]$$

$$\text{or,} \quad \frac{2}{3} \dot{u} + \frac{1}{3} (\dot{Du})D^{-1} = \frac{2}{3} \left((D^3 u) + u(Du)\right)$$

$$+ \frac{1}{3} \left(D((D^3 u) + u(Du))\right)D^{-1}$$

Thus comparing the terms on both sides of the equation, we obtain

$$\dot{u} = (D^3 u) + u(Du) \tag{9.70}$$

which is the KdV equation. Thus we see that with a little care, the geometrical approach extends easily to the continuum models also.

References:

Crampin, M., G. Marmo and C. Rubano, Phys. Lett. 97A, 88 (1983).

Curtis, W. D. and F. R. Miller, Differential Manifolds
        and Theoretical Physics, Academic Press, 1985.

Das, A. and S. Okubo, Ann. Phys. <u>190</u> (1989).

Kobayashi, S. and K. Nomizu, Foundations of Differential
        Geometry, Vol. I, Interscience, 1963.

Magri, F., J. Math. Phys. <u>19</u>, 1156 (1978).

Marmo, G. in Proceedings of the International Meeting on
        Geometry and Physics, Florence, 1982, Ed. M.
        Modugno.

Nijenhuis, A., Indag. Math. <u>13</u>, 200 (1951).

Okubo, S. and A. Das, Phys. Lett. <u>B209</u>, 311 (1988).

Olver, P. J., J. Math. Phys. <u>18</u>, 1212 (1977).

CHAPTER 10

THE TODA LATTICE

The main example we have studied so far, namely, the Korteweg-de Vries equation, is a continuum model. In many respects, a finite dimensional system with a finite number of degrees of freedom is simpler to study. And furthermore, since the geometrical approach to integrable models, which we presented in the last chapter, was primarily described for a finite dimensional system, study of such a system seems appropriate at this point. The Toda lattice is such a system and in this chapter we will study this system and its integrability from a geometrical point of view deferring a group theoretical treatment till the next chapter.

The Toda Equation:

The Toda lattice is a finite dimensional system and describes the motion of N point masses on the line under the influence of an exponential interaction. The Hamiltonian equation in terms of the canonical coordinates Q_i and momenta $P_i (i=1,2,\ldots,N)$ are given by

$$\dot{Q}_i = P_i \qquad\qquad i = 1,2\ldots,N$$

$$\dot{P}_j = e^{-(Q_j-Q_{j-1})} - e^{-(Q_{j+1}-Q_j)} \qquad j = 2,3,\ldots,N-1$$

$$\dot{P}_1 = - e^{-(Q_2-Q_1)} \qquad\qquad\qquad (10.1)$$

$$\dot{P}_N = e^{-(Q_N-Q_{N-1})}$$

It is clear that the evolution equations for the momenta are
not quite symmetrical. However, we can cast the equations
into a more symmetrical form by enlarging the system to
consist of N+2 point masses with the end points held fixed
at spatial infinity, namely, if we define

$$Q_0 = -\infty$$

$$\text{(10.2)}$$

$$Q_{N+1} = \infty$$

then the Hamiltonian equations in (10.1) take the more
symmetrical form

$$\dot{Q}_i = P_i$$

$$\text{(10.3)}$$

$$\dot{P}_i = e^{-(Q_i - Q_{i-1})} - e^{-(Q_{i+1} - Q_i)} \qquad i = 1, 2, \ldots, N$$

In what follows, we would assume the boundary conditions of
Eq. (10.2) and the more symmetrical form of the Toda
equations given above.   Let us also note here that the
canonical coordinates can simply   be thought of as a
particular choice of the more generalized coordinates, y^μ,
of the phase space discussed earlier such that

$$y^i = Q_i$$

$$\text{(10.4)}$$

$$y^{N+i} = P_i \qquad\qquad i = 1, 2 \ldots, N$$

Dual Poisson Bracket Structure:

The next step in the geometrical approach consists of finding two inequivalent Lagrangians which give Eq. (10.3) as their Euler-Lagrange equations. The canonical first order Lagrangian is easily obtained to be

$$L_o = \sum_{\mu=1}^{2N} \theta_\mu^{(o)}(y)\dot{y}^\mu - H_o(y)$$

$$= \sum_{i=1}^{N} \frac{1}{2} (P_i\dot{Q}_i - Q_i\dot{P}_i) - H_o(Q,P) \qquad (10.5)$$

where

$$H_o(Q,P) = \sum_{i=1}^{N} \left(\frac{1}{2} P_i^2 + e^{-(Q_{i+1}-Q_i)}\right) \qquad (10.6)$$

Note that we have introduced in this chapter the explicit summation to avoid any possible confusion. The symplectic structure in the present case can be worked out simply and takes the matrix form in terms of N×N blocks as

$$f_{\mu\nu}(y) = \partial_\mu \theta_\nu^{(o)}(y) - \partial_\nu \theta_\mu^{(o)}(y) = \left(\begin{array}{c|c} 0 & -I \\ \hline I & 0 \end{array}\right) \qquad (10.7)$$

This is indeed the canonical symplectic structure of Eq. (9.23) which gives the canonical symplectic form of the manifold, namely,

$$f = \frac{1}{2} \sum_{\mu,\nu=1}^{2N} f_{\mu\nu}(y) dy^\mu \wedge dy^\nu$$

$$= - \sum_{i=1}^{N} dQ_i \wedge dP_i \tag{10.8}$$

The canonical Poisson bracket structure $f^{\mu\nu}(y)$ is obtained from Eq. (10.7) as the inverse so that

$$f^{\mu\nu}(y) = \left(\begin{array}{c|c} 0 & I \\ \hline -I & 0 \end{array} \right) \tag{10.9}$$

A second Lagrangian whose Euler-Lagrange equations also give the same equations as those of Eq. (10.3) can be determined after some work to be

$$L = \sum_{\mu=1}^{2N} \theta_\mu(y) \dot{y}^\mu - H(y)$$

$$= \sum_{i=1}^{N} \left\{ \left(\frac{1}{2} P_i^2 + e^{-(Q_{i+1}-Q_i)} \right) \dot{Q}_i + \pi_i(P) \dot{P}_i \right\} - H(Q,P) \tag{10.10}$$

where

$$\pi_i(P) = \frac{1}{2} \sum_{j=1}^{N} \epsilon(i-j) P_j$$

$$H(Q,P) = \sum_{i=1}^{N} \left(\frac{1}{3} P_i^3 + (P_i + P_{i+1}) e^{-(Q_{i+1}-Q_i)} \right) \tag{10.11}$$

Here we have also used the alternating function defined to be

$$\epsilon(i-j) = \} \begin{matrix} 1 \\ 0 \\ -1 \end{matrix} \quad \begin{matrix} \text{if} \\ \text{if} \\ \text{if} \end{matrix} \quad \begin{matrix} i>j \\ i=j \\ i<j \end{matrix} \qquad (10.12)$$

A direct verification of the fact that the Euler-Lagrange equations following from Eq. (10.10) coincide with those of Eq. (10.3) is rather tedious but straightforward. One could also check this indirectly as follows. We can calculate the symplectic structure associated with the Lagrangian of Eq. (10.10) which takes the matrix form in terms of N×N blocks as

$$F_{\mu\nu}(y) = \partial_\mu \theta_\nu(y) - \partial_\nu \theta_\mu(y)$$

$$= \left(\begin{array}{c|c} A & -B \\ \hline B & e \end{array} \right) \qquad (10.13)$$

where the block matrices have the explicit form

$$A_{ij} = \delta_{i+1,j} \, e^{-(Q_{i+1}-Q_i)} - \delta_{i,j+1} \, e^{-(Q_{j+1}-Q_j)}$$

$$B_{ij} = P_i \, \delta_{i,j} \qquad\qquad (10.14)$$

$$e_{ij} = \epsilon(j-i)$$

The symplectic form then follows to be

$$F = \frac{1}{2} \sum_{\mu,\nu=1}^{2N} F_{\mu\nu}(y) dy^\mu \wedge dy^\nu$$

$$= \sum_{i=1}^{N} \left[e^{-(Q_{i+1}-Q_i)} dQ_i \wedge dQ_{i+1} - P_i dQ_i \wedge dP_i \right]$$

$$+ \frac{1}{2} \sum_{i,j=1}^{N} \epsilon(j-i) dP_i \wedge dP_j \qquad (10.15)$$

Let us next define the (1,1) tensor

$$S_\mu{}^\nu(y) = \sum_{\lambda=1}^{2N} F_{\mu\lambda}(y) f^{\lambda\nu}(y)$$

which takes the matrix form

$$S_\mu{}^\nu = \left(\begin{array}{c|c} B & A \\ \hline -e & B \end{array} \right) \qquad (10.16)$$

We recognize from our discussion of Eq. (9.26) that the Euler-Lagrange equations following from Eq. (10.5) and (10.10) would be the same if

$$\sum_{\nu=1}^{2N} F^{\mu\nu}(y) \partial_\nu H(y) = \sum_{\nu=1}^{2N} f^{\mu\nu}(y) \partial_\nu H_o(y)$$

$$\text{or,} \quad \partial_\mu H(y) = \sum_{\lambda,\nu=1}^{2N} F_{\mu\lambda}(y) f^{\lambda\nu}(y) \partial_\nu H_o(y)$$

$$\text{or,} \quad \partial_\mu H(y) = \sum_{\nu=1}^{2N} S_\mu{}^\nu(y) \partial_\nu H_o(y) \qquad (10.17)$$

That this is satisfied for the present case can be checked from the forms of $H_o(Q,P)$, $H(Q,P)$ and $S_\mu^\nu(y)$ given in Eqs. (10.6), (10.11) and (10.16) respectively. We have, therefore, shown that the Toda equations are compatible with a dual symplectic structure. We have constructed the two distinct symplectic structures for the Toda system as well as the (1,1) tensor, $S_\mu^\nu(y)$, associated with them. It is easy to see now how the discussion of the geometrical approach in Chapter 9 applies to this system.

Conserved Quantities:

Given $S_\mu^\nu(y)$ - the (1,1) tensor constructed from the two symplectic structures - we can write down the conserved quantities for the system following our earlier discussion. As we have seen in Eq. (9.43), these are given by $\mathrm{Tr}S^n$ for various powers n. Let us derive the first few for completeness.

$$\mathrm{Tr}S = 2\mathrm{Tr}B = 2 \sum_{i=1}^{N} P_i$$

Thus

$$H_1 = \frac{1}{2} K_1 = \frac{1}{2} \mathrm{Tr}S = \sum_{i=1}^{N} P_i \qquad (10.18)$$

is the first conserved quantity. We recognize this as the

total momentum of the system and the conservation is a consequence of the translation invariance of the system. Next we note that

$$\mathrm{Tr}S^2 = \mathrm{Tr}\left(2B^2 - (Ae + eA)\right)$$

Upon simplification, this gives

$$H_2 = \tfrac{1}{2} K_2 = \tfrac{1}{4} \mathrm{Tr}S^2 = \sum_{i=1}^{N} \left[\tfrac{1}{2} P_i^2 + e^{-(Q_{i+1} - Q_i)}\right]$$

$$= H_o(Q, P) \qquad (10.19)$$

This is, of course, the Hamiltonian associated with the first symplectic structure and can, consequently, be thought of as the energy of the canonical system. Let us simply write down the next two conserved quantities below.

$$H_3 = \tfrac{1}{2} K_3 = \tfrac{1}{6} \mathrm{Tr}S^3$$

$$= \sum_{i=1}^{N} \left[\tfrac{1}{3} P_i^3 + (P_i + P_{i+1}) e^{-(Q_{i+1} - Q_i)}\right]$$

$$= H(Q, P) \qquad (10.20)$$

and

$$H_4 = \tfrac{1}{2} K_4 = \tfrac{1}{8} \mathrm{Tr}S^4$$

$$= \sum_{i=1}^{N} \left[\tfrac{1}{4} P_i^4 + \left(P_i^2 + P_i P_{i+1} + P_{i+1}^2\right) e^{-(Q_{i+1} - Q_i)}\right.$$

$$\left. + \tfrac{1}{2} e^{-2(Q_{i+1} - Q_i)} + e^{-(Q_{i+2} - Q_i)}\right] \quad (10.21)$$

It is clear that the conserved quantities $K_n = \frac{1}{2n} \mathrm{Tr} S^n$ would involve, among other things, terms proportional to $\mathrm{Tr} B^n$. These would give rise to the pure momentum dependent terms of the conserved quantities which would have the form $\sum_{i=1}^{N} P_i^n$. It can be seen from simple algebra that given N independent quantities $P_1, P_2, \ldots, P_N$, the following algebraic combinations

$$\sum_{i=1}^{N} P_i^n \qquad\qquad n = 1, 2, \ldots N$$

are independent. Consequently, we conclude that of all the conserved quantities, only N, namely, $H_1, H_2 \ldots H_N$ are functionally independent. This is precisely the right number of conserved quantities needed to prove that the Toda system is integrable. We, of course, must also prove that these conserved quantities are in involution which we do next.

The Nijenhuis Tensor:

As we have seen in the last chapter (see Eqs. (9.48)-(9.50)), the vanishing of the Nijenhuis tensor associated with $S_\mu^{\ \nu}(y)$ leads to the involution of the conserved quantities constructed earlier. In the present case the form of $S_\mu^{\ \nu}(y)$ is determined in Eq. (10.16) and consequently, the explicit form of the Nijenhuis tensor can be obtained from the definition

$$
N^{\mu}_{\alpha\beta} = \sum_{\lambda=1}^{2N} \left[s_{\alpha}{}^{\lambda} \partial_{\lambda} s_{\beta}{}^{\mu} - s_{\beta}{}^{\lambda} \partial_{\lambda} s_{\alpha}{}^{\mu} \right.
$$

$$
\left. - s_{\lambda}{}^{\mu} \left(\partial_{\alpha} s_{\beta}{}^{\lambda} - \partial_{\beta} s_{\alpha}{}^{\lambda} \right) \right] \quad (10.22)
$$

If we further note that by definition

$$
N^{\mu}_{\alpha\beta} = -N^{\mu}_{\beta\alpha} \quad\quad\quad\quad (10.23)
$$

then the vanishing of the Nijenhuis tensor in the present case can be systematically seen in the following way.

Let us first examine the components $N^{i}_{\alpha\beta}$ with $i=1,2,\ldots N$ and $\alpha,\beta = 1,2,\ldots 2N$. From Eq. (10.22) we can easily work out to see that

$$
N^{i}_{\alpha\beta} = \delta^{i}_{\beta} \left(\delta^{i}_{\alpha+1} e^{-(Q_{\alpha+1}-Q_{\alpha})} - \delta^{i}_{\alpha-1} e^{-(Q_{\alpha}-Q_{\alpha-1})} \right)
$$

$$
- \delta^{i}_{\alpha} \left(\delta^{i}_{\beta+1} e^{-(Q_{\beta+1}-Q_{\beta})} - \delta^{i}_{\beta-1} e^{-(Q_{\beta}-Q_{\beta-1})} \right)
$$

$$
+ \delta_{\alpha,\beta+1} \left(\epsilon(\beta-i+1) - \epsilon(\beta-i) \right) e^{-(Q_{\beta+1}-Q_{\beta})}
$$

$$
- \delta_{\alpha+1,\beta} \left(\epsilon(\alpha-i+1) - \epsilon(\alpha-i) \right) e^{-(Q_{\alpha+1}-Q_{\alpha})} \quad (10.24)
$$

If we now use the identity

$$
\epsilon(\alpha-i+1) - \epsilon(\alpha-i) = \delta_{\alpha,i} + \delta_{\alpha,i-1} \quad\quad (10.25)
$$

then Eq. (10.24) gives

$$N^i_{\alpha\beta} = 0 \qquad\qquad i = 1,2,\ldots.N \qquad\qquad (10.26)$$

$$\alpha,\beta = 1,2,\ldots.2N$$

Next let us look at the components $N^{i+N}_{j+N,k+N}$ with $i,j,k = 1,2\ldots.N$. From Eq. (10.22) we see that this takes the simple form

$$N^{i+N}_{j+N,k+N} = P_i\delta^i_j\delta_{jk} - P_i\delta^i_k\delta^i_j$$

$$- P_i\delta^i_j\delta_{jk} + P_i\delta^i_k\delta^i_j = 0 \qquad\qquad (10.27)$$

Similarly, the components $N^{i+N}_{j+N,k}$ simplify after a little calculation to the form

$$N^{i+N}_{j+N,k} = \delta^i_{k+1}\big(\epsilon(j-k)-\epsilon(j-k-1)\big)e^{-(Q_{k+1}-Q_k)}$$

$$+ \delta^i_{k-1}\big(\epsilon(j-k)-\epsilon(j-k+1)\big)e^{-(Q_k-Q_{k-1})}$$

$$- \delta^i_j\big(\delta^i_{k+1}e^{-(Q_{k+1}-Q_k)} - \delta^i_{k-1}e^{-(Q_k-Q_{k-1})}\big)$$

$$- \delta_{j,k}\big(\delta^i_{j+1}e^{-(Q_{j+1}-Q_j)} - \delta^i_{j-1}e^{-(Q_j-Q_{j-1})}\big) \qquad (10.28)$$

Upon using the identity in Eq. (10.25), we obtain

$$N_{j+N,k}^{i+N} = 0 \qquad\qquad i,j,k = 1,2\ldots.N \qquad\qquad (10.29)$$

Finally, the components $N_{j,k}^{i+N}$ have the form

$$N_{j,k}^{i+N} = (P_j - P_i) \frac{\partial}{\partial Q_j} \left(\delta_{k+1}^{i} e^{-(Q_i - Q_{i-1})} - \delta_{k-1}^{i} e^{-(Q_{i+1} - Q_i)} \right)$$

$$- (P_k - P_i) \frac{\partial}{\partial Q_k} \left(\delta_{j+1}^{i} e^{-(Q_i - Q_{i-1})} - \delta_{j-1}^{i} e^{-(Q_{i+1} - Q_i)} \right)$$

$$(10.30)$$

After expanding this out, it is seen to vanish, namely,

$$N_{j,k}^{i+N} = 0 \qquad\qquad\qquad\qquad (10.31)$$

Combining the results of Eqs. (10.26), (10.27), (10.29) and (10.31) as well as the antisymmetry property of Eq. (10.23), we conclude that for the Toda lattice

$$N_{\alpha\beta}^{\mu} = 0 \qquad\qquad \mu,\alpha,\beta = 1,2,\ldots.2N \qquad\qquad (10.32)$$

As we have seen in Eqs. (9.48)-(9.50), the vanishing of the Nijenhuis tensor induces a recursion relation between the conserved quantities

$$H_n = \frac{1}{2n} \, Trs^n \qquad\qquad n = 1,2,\ldots.N \qquad\qquad (10.33)$$

In the present case, the recursion relation takes the form

$$\partial_\mu H_{n+1} = \sum_{\nu=1}^{2N} s_\mu{}^\nu \partial_\nu H_n \qquad (10.34)$$

with $s_\mu{}^\nu(y)$ defined in Eq. (10.16). The recursion relation leads to the involution of the conserved quantities, namely,

$$\{H_n, H_m\}_0 = 0 = \{H_n, H_m\} \qquad (10.35)$$

Thus we have shown that the Toda lattice not only has the right number of conserved quantities to be integrable, but the conserved quantities are also in involution so that the Toda lattice indeed is an integrable system. Let us also note here that the higher order equations of the Toda hierarchy are given by

$$\dot{y}^\mu = f^{\mu\nu} \partial_\nu H_n = F^{\mu\nu} \partial_\nu H_{n+1} \qquad (10.36)$$

and share the same conserved quantities and, consequently, are also integrable.

The Lax Equation:

We have seen earlier that the Lax equation plays an important role in the study of integrable models. Furthermore, we know that in the geometrical approach, we have a systematic method of constructing the Lax operators

and the Lax equation for the entire hierarchy of equations.
In what follows, we will construct the Lax equations for the
Toda hierarchy from the geometrical approach.

Let us recall from Eqs. (9.37) and (9.41) that the Lax
pair consists of two operators one of which is $S_\mu{}^\nu(y)$. The
second operator is related to the equation of motion and for
the Toda equations (10.3) is defined to be

$$U_\mu{}^\nu(y) = \partial_\mu\left(f^{\nu\lambda}\partial_\lambda H_2(y)\right) = \partial_\mu\left(f^{\nu\lambda}\partial_\lambda H_0(y)\right) \qquad (10.37)$$

Here we have used the identification given in Eq. (10.19).
Since $f^{\mu\nu}$ is the canonical Poisson bracket structure and is,
therefore, constant, we have

$$U_\mu{}^\nu(y) = f^{\nu\lambda}\partial_\mu\partial_\lambda H_0(y) \qquad (10.38)$$

Given the form of. H_0 in Eq. (10.6), we can easily calculate
this. In terms of N×N blocks, this takes the form

$$U_\mu{}^\nu = \left(\begin{array}{c|c} 0 & -D \\ \hline I & 0 \end{array}\right) \qquad (10.39)$$

where I is the identity matrix and

$$D_i{}^j = \left(\delta_i^j - \delta_{i+1}^j\right)e^{-(Q_{i+1}-Q_i)}$$

$$+ \left(\delta_i^j - \delta_{i-1}^j\right)e^{-(Q_i-Q_{i-1})} \qquad (10.40)$$

Thus the Lax representation for the Toda equations, namely,

$$\frac{dS}{dt} = [S, U]$$

takes the matrix form

$$\left(\begin{array}{c|c} \dfrac{dB}{dt} & \dfrac{dA}{dt} \\ \hline -\dfrac{de}{dt} & \dfrac{dB}{dt} \end{array}\right) = \left(\begin{array}{c|c} A-De & -[B,D] \\ \hline 0 & eD-A \end{array}\right) \qquad (10.41)$$

Thus the Lax equation in the present case reduces to the following N×N matrix equations

$$\frac{de}{dt} = 0 \qquad\qquad (10.42)$$

$$\frac{dA}{dt} = -[B, D] \qquad\qquad (10.43)$$

$$\frac{dB}{dt} = A - De = eD - A = \frac{1}{2}[e, D] \qquad\qquad (10.44)$$

We see that Eq. (10.42) is a consistency condition since e is a constant matrix. Therefore, there are only two independent matrix equations. If we work out in detail, Eqs. (10.43) and (10.44) give respectively

$$\dot{Q}_i = P_i$$

$$\dot{P}_i = e^{-(Q_i - Q_{i-1})} - e^{-(Q_{i+1} - Q_i)}$$

These are nothing other than the Toda equations of (10.3).

We can, of course, also construct the Lax representations for the higher order equations of the hierarchy. In this case, the second Lax operator is simply given by

$$(U_n)_\mu{}^\nu = \partial_\mu\!\left(f^{\nu\lambda}\partial_\lambda H_n\right) = f^{\nu\lambda}\partial_\mu\partial_\lambda H_n \qquad\qquad (10.45)$$

which can be worked out explicitly since we know the exact forms of H_n. The Lax representation for the nth order equation, then, simply becomes

$$\frac{dS}{dt_n} = [S,U_n] \qquad\qquad (10.46)$$

Several comments are in order here. Our construction of the Lax operators has followed the geometrical approach and consequently, has a built in symplectic character. For example, the Lax pairs consist of 2N×2N matrices. In fact, even the nontrivial N×N matrix equations consist of a diagonal matrix equation, namely Eq. (10.44), and an antisymmetric matrix equation, namely Eq. (10.43). The Toda system, of course, has been studied from various points of view and needless to say that other Lax representations of the system (not from the geometrical approach) exist. It would be instructive to compare some of these. For example, the conventional Lax pair of Flaschka and Moser are given in terms of N×N matrices and have the form

$$\tilde{S}_i{}^j = \frac{1}{2}\left(P_i\delta_i^j + \delta_{i+1}^j e^{-(Q_{i+1}-Q_i)/2}\right.$$

$$\left. + \delta_i^{j+1} e^{-(Q_{j+1}-Q_j)/2}\right) \qquad (10.47)$$

$$\tilde{U}_i{}^j = \frac{1}{2}\left(\delta_{i+1}^j e^{-(Q_{i+1}-Q_i)/2}\right.$$

$$\left. - \delta_i^{j+1} e^{-(Q_{j+1}-Q_j)/2}\right) \qquad (10.48)$$

The Lax equation

$$\frac{d\tilde{S}}{dt} = [\tilde{S},\tilde{U}] \qquad (10.49)$$

of course, has the same content as Eq. (10.3). Let us also note that the Lax operator $\tilde{S}$ can be separated into a diagonal and an off-diagonal part as

$$\tilde{S}_i{}^j = \tilde{B}_i{}^j + \tilde{A}_i{}^j \qquad (10.50)$$

where

$$\tilde{B}_i{}^j = \frac{1}{2} P_i\delta_i^j \qquad (10.51)$$

$$\tilde{A}_i{}^j = \frac{1}{2}\left(\delta_{i+1}^j e^{-(Q_{i+1}-Q_i)/2} + \delta_i^{j+1} e^{-(Q_{j+1}-Q_j)/2}\right) \qquad (10.52)$$

That is, the diagonal part, $\tilde{B}$, coincides up to a

multiplicative constant with the B matrix we have
constructed earlier. The off-diagonal matrix, $\tilde{A}$, on the
other hand, is quite different from the A matrix even though
the structures look similar. For example, $\tilde{A}$ is a symmetric
matrix whereas A is antisymmetric.

With the decomposition in Eq. (10.50), the conventional
Lax equation of (10.49) also separates into two NxN matrix
equations given by

$$\frac{d\tilde{A}}{dt} = [\tilde{B}, \tilde{U}] \qquad\qquad (10.53)$$

$$\frac{d\tilde{B}}{dt} = [\tilde{A}, \tilde{U}] \qquad\qquad (10.54)$$

While the contents of Eqs. (10.53) and (10.54) are the same
as those of Eqs. (10.43) and (10.44), the structures are
very different. In fact, Eq. (10.53) is a symmetric matrix
equation whereas Eq. (10.43) is in terms of an antisymmetric
matrix. Furthermore, since the similarity transformations
preserve the symmetry properties of a matrix, the two Lax
representations are indeed inequivalent.

References:

Das, A. and S. Okubo, Ann. Phys. 190 (1989).

Flaschka, H., Phys. Rev. B9, 1424 (1974).

Flaschka, H., Prog. Theor. Phys. 51, 703 (1974).

Flaschka, H. in Dynamical Systems, Theory and Applica-
        tions, Ed. J. Moser, Springer-Verlag, 1974.

Hénon, M., Phys. Rev. B9, 1421 (1974).

Moser, J. in Dynamical Systems, Theory and Applications, Ed. J. Moser, Springer-Verlag, 1974.

Toda, M., Prog. Theor. Phys. Suppl. $\underline{59}$, 1 (1976).

Toda, M., Theory of Nonlinear Lattices, Springer-Verlag, 1981.

CHAPTER 11

GROUP THEORETICAL APPROACH TO THE TODA LATTICE

In the last chapter, we studied the Toda lattice from a geometrical approach. We explicitly constructed the conserved quantities and demonstrated the integrability of the system following from the vanishing of the Nijenhuis torsion tensor. In this chapter, we will examine the integrability of the Toda lattice from a group theoretical point of view. This has the advantage that it emphasizes the important role played by the symmetry groups in the integrability of a system. Furthermore, the ideas developed in this chapter become quite useful in the study of the Yang-Baxter equation as well as the quantum inverse scattering method to be discussed later.

Review of Lie Algebra:

The most important part in the study of a continuous symmetry group is its Lie algebra. There exist many excellent books on the study of Lie algebra and one should refer to these for derivation of various results on the subject. Here we would merely summarize the results which would prove useful in our study. Furthermore, although the results would be quite general, wherever possible we would give SU(N) as an example since that is the group which would enter our discussion of the Toda lattice.

Let us consider a continuous simple group G and denote its Lie algebra by g. The Lie algebra consists of the elements, T_I, with $I = 1,2,....$dim G and the multiplication rule

$$[T_I,T_J] = if^K_{IJ}T_K \qquad I,J,K = 1,2,...\text{dim G} \quad (11.1)$$

Here we have defined the elements of g to be Hermitian and f^K_{IJ} are the structure constants of the group. The elements, T_I, generate the group G and are, in fact, defined to be the group elements near the identity element. Geometrically speaking, therefore, if one considers a Lie group as a manifold, then its Lie algebra simply corresponds to the tangent space at the identity and the generators, T_I, merely constitute a basis of the tangent manifold. It is obvious from this also that dim G represents the dimensionality of the group manifold. Namely, it corresponds to the number of independent basis elements needed to span the manifold. In the language of matrices, dim G represents the number of independent matrices needed to specify any group element.

Since the T_I's constitute a basis of the simple Lie algebra, we could choose them to be orthonormal, satisfying in a matrix representation the condition

$$\text{Tr } T_I T_J = \delta_{IJ} \qquad (11.2)$$

A second inherent characteristic of a Lie group G is its rank denoted by rank G. This is defined to be the maximum number of commuting elements of the Lie algebra in Eq. (11.1). In the matrix representation, rank G denotes the maximum number of generators, T_I, which can be diagonalized. Let us note here that for the group SU(N), which is defined by the properties of the $N \times N$ unitary matrices with determinant unity,

$$\dim SU(N) = N^2 - 1$$

and   (11.3)

$$\text{rank } SU(N) = N-1$$

Although the orthonormal basis of Eq. (11.2) is the natural basis for the Lie algebra, various algebraic questions such as the representation theory are much better studied in an alternate basis known as the Cartan-Weyl basis. Here the generators are divided into two classes. The first consists of the commuting elements of the simple Lie algebra, denoted by H_i with $i = 1,2,....$rank G, giving the maximally commuting subalgebra of the Lie algebra, namely,

$$[H_i, H_j] = 0 \qquad\qquad i,j = 1,2...\text{rank G} \quad (11.4)$$

This is also called the Cartan subalgebra of the Lie algebra and the generators H_i are defined to be hermitian. The second set denoted by e_α, is constructed by taking complex combinations of the remaining elements of the Lie algebra such that

$$[H_i, e_\alpha] = \alpha_i e_\alpha \qquad\qquad i = 1,2...\text{rank G} \quad (11.5)$$

Here α_i's are the structure constants in the Cartan-Weyl basis and correspond to the components of a Euclidean vector of dimensionality rank G, for each generator e_α. These vectors are known as the root vectors and label the generators which do not belong to the Cartan subalgebra. Clearly, the number of roots for a Lie algebra is given by

$$\text{Number of roots} = \dim G - \text{rank } G \qquad (11.6)$$

For SU(N), this number is

$$\text{Number of roots} = N^2 - 1 - (N-1) = N(N-1) \qquad (11.7)$$

The roots can be divided into positive and negative roots in the following way. If the first nonzero component of a root vector $\alpha = \{\alpha_i\}$ is positive, then it is defined to be a positive root. It is negative otherwise. Since α_i's are real, we obtain from Eq. (11.5) that

$$[H_i, e_\alpha]^+ = \alpha_i e_\alpha^+$$

$$\text{or,} \quad [H_i, e_\alpha^+] = -\alpha_i e_\alpha^+ \qquad (11.8)$$

This shows that if α is a root vector, then so is $(-\alpha)$ and that

$$e_{-\alpha} = e_\alpha^+ \qquad (11.9)$$

Finally, the commutation relations between the e_α's can be obtained from the Jacobi identity and take the form

$$[e_\alpha, e_\beta] = \alpha_i H_i \qquad\qquad \text{if } \alpha+\beta = 0$$

$$= N(\alpha,\beta)e_{\alpha+\beta} \qquad \text{if } \alpha+\beta \text{ is a root} \quad (11.10)$$

$$= 0 \qquad\qquad\qquad \text{otherwise}$$

The constants $N(\alpha,\beta)$ can be chosen to be real and can be calculated explicitly if we know all the roots. However, the $N(\alpha,\beta)$'s, in general, satisfy various identities which are often quite useful. For example, let us note without derivation (The derivation is actually quite straight-forward.) that

$$N(\alpha,\beta) = -N(\beta,\alpha) = -N(-\alpha,-\beta) \qquad\qquad (11.11)$$

$$N(\alpha,\beta) = N(-\alpha,\alpha+\beta) = -N(-\beta,\alpha+\beta) \qquad\qquad (11.12)$$

Combining Eqs. (11.11) and (11.12) we obtain

$$N(\alpha,\beta) = -N(-\alpha-\beta,\beta) \qquad\qquad (11.13)$$

The structure of a Lie group is contained in its structure constants and as we have noted earlier these are nothing other than the root vectors of the Lie algebra. Furthermore, since the negative roots are simply negative of the positive roots, the structure is essentially contained in the set of positive root vectors. The root structure can be further simplified by noting that the positive roots, in general, are not all linearly independent. In fact, the number of linearly independent positive roots can only equal rank G since that is the dimensionality of the space in

which they live.   Let   us   next   define   a simple root as a
positive root which   cannot   be   written   as   the sum of two
positive roots.   With a little   bit of work,   it can be shown
that the number of simple roots   is rank G and that they are
linearly independent.   Furthermore,   if   we denote a simple
root by

$$\alpha^a \qquad\qquad\qquad\qquad a = 1,2....\text{rank } G \quad (11.14)$$

then any positive root can be written as

$$\alpha = \sum_{a=1}^{\text{rank } G} k_a \alpha^a \qquad\qquad\qquad (11.15)$$

where k_a's   are   non-negative   integers.   Consequently, the
information about the   structure   of   the   group now resides
completely in the set of   simple   roots.   In fact, given two
simple roots α^a and α^b, the Cartan matrix is defined as

$$K_{ab} = \frac{2(\alpha^a, \alpha^b)}{(\alpha^b, \alpha^b)} \qquad\qquad\qquad (11.16)$$

where (,)   stands   for   the   Euclidean inner product.   The
Cartan matrix encodes all   the   information about the group.
Let us simply note here for   future use that, for SU(N), the
Cartan matrix is a (N-1)×(N-1) matrix of the form

$$K_{ab} = 2\delta_{a,b} - \delta_{a+1,b} - \delta_{a,b+1}$$

$$
\text{or, } K_{ab} = \begin{pmatrix}
2 & -1 & 0 & \cdots\cdots\cdots\cdots\cdots \\
-1 & 2 & -1 & 0 & \cdots\cdots\cdots \\
0 & -1 & 2 & -1 & 0 & \cdots\cdots \\
\cdot & & & & & & \\
\cdot & & & & & & \\
\cdot & & & 0 & -1 & 2 & -1 \\
0 & \cdots\cdots\cdots\cdots & 0 & -1 & 2
\end{pmatrix} \qquad (11.17)
$$

The inverse K_{ab}^{-1} is a full matrix but satisfies the identity

$$
\sum_c K_{ac}^{-1} K_{cb} = \delta_{ab}
$$

Using the formula in Eq. (11.17) for K_{ab}, this leads to

$$
2K_{ab}^{-1} - K_{a\,b-1}^{-1} - K_{a\,b+1}^{-1} = \delta_{ab} \qquad (11.18)
$$

Since the orthonormal basis and the Cartan-Weyl basis describe the same Lie algebra, they must be unitarily related. That is, if we write

$$
T_I = \sum_{i=1}^{\text{rank}G} c_{Ii} H_i + \sum_{\substack{\alpha > 0 \\ \text{roots}}} \left(c_{I\alpha} e_\alpha + c_{I-\alpha} e_{-\alpha} \right) \qquad (11.19)
$$

then hermiticity would require

$$
c_{Ii}^* = c_{Ii}
$$

$$
c_{I\alpha}^* = c_{I-\alpha}
$$

$$(11.20)$$

and unitarity would lead to the relation

$$\sum_{i=1}^{\text{rank } G} c_{Ii}c_{Ji} + \sum_{\substack{\alpha > 0 \\ \text{roots}}} \left(c_{I\alpha}c_{J-\alpha} + c_{I-\alpha}c_{J\alpha} \right) = \delta_{IJ} \qquad (11.21)$$

We note here that a second basis that is sometimes used in the algebraic studies is the Chevalley basis. Here one defines

$$H_\alpha = \frac{2\alpha_i H_i}{(\alpha,\alpha)}$$

$$(11.22)$$

$$E_\alpha = \sqrt{\frac{2}{(\alpha,\alpha)}} \, e_\alpha$$

The form of the commutation relations in this case are obtained from Eqs. (11.5) and (11.10) and are given by

$$[H_\alpha, H_\beta] = 0$$

$$[H_\alpha, E_\beta] = E_\beta \frac{2(\beta,\alpha)}{(\alpha,\alpha)} \qquad (11.23)$$

$$[E_\alpha, E_\beta] = \sqrt{\frac{2(\alpha+\beta,\alpha+\beta)}{(\alpha,\alpha)(\beta,\beta)}} \, N(\alpha,\beta) E_{\alpha+\beta} \quad \text{if} \quad \alpha+\beta \text{ is a root}$$

$$= H_\alpha \qquad\qquad\qquad\qquad \text{if} \quad \alpha+\beta = 0$$

$$= 0 \qquad\qquad\qquad\qquad\qquad \text{otherwise}$$

Note in particular that for

$$H_a = H_{\alpha^a}$$

and   (11.24)

$$E_a = E_{\alpha^a}$$

the commutation relations take the simple form

$$[H_a, H_b] = 0$$

$$[H_a, E_b] = E_b K_{ba}$$

                                                                       (11.25)

$$[E_a, E_{-a}] = H_a$$

$$[E_a, E_b] = \sqrt{\frac{2(\alpha^a + \alpha^b, \alpha^a + \alpha^b)}{(\alpha^a, \alpha^a)(\alpha^b, \alpha^b)}} \; N(\alpha^a, \alpha^b) E_{\alpha^a + \alpha^b}$$

Group structure of the Toda equations:

Let us next try to identify the group structure associated with the Toda lattice. We recall from Eq. (10.1) that in terms of N canonical coordinates Q_i and momenta P_i, the Toda equations take the form

$$\dot{Q}_i = \dot{P}_i \qquad\qquad i = 1, 2 \ldots N \quad (11.26)$$

$$\dot{P}_j = e^{-(Q_j - Q_{j-1})} - e^{-(Q_{j+1} - Q_j)} \qquad j = 2, 3, \ldots N-1$$

$$\dot{P}_1 = -e^{-(Q_2 - Q_1)}$$

                                                                       (11.27)

$$\dot{P}_N = e^{-(Q_N - Q_{N-1})}$$

We can write this equivalently as the following set of second order equations

$$\ddot{Q}_1 = \dot{P}_1 = -e^{-(Q_2-Q_1)}$$

$$\ddot{Q}_2 = \dot{P}_2 = e^{-(Q_2-Q_1)} - e^{-(Q_3-Q_2)}$$

$$\ddot{Q}_3 = \dot{P}_3 = e^{-(Q_3-Q_2)} - e^{-(Q_4-Q_3)}$$

.

.   (11.28)

.

.

.

$$\ddot{Q}_{N-1} = \dot{P}_{N-1} = e^{-(Q_{N-1}-Q_{N-2})} - e^{-(Q_N-Q_{N-1})}$$

$$\ddot{Q}_N = \dot{P}_N = e^{-(Q_N-Q_{N-1})}$$

It is obvious from Eq. (11.28) that

$$\sum_{i=1}^{N} \ddot{Q}_i = \sum_{i=1}^{N} \dot{P}_i = 0$$

Namely, the total momentum is constant and, therefore, the center of mass motion can be separated and the dynamics of the system can be written in terms of (N-1) coordinates and momenta. In fact, let us define the (N-1) coordinates as

$$q_a = Q_{a+1} - Q_a \qquad\qquad a = 1,2,....N-1 \qquad (11.29)$$

The second order equations satisfied by the q_a's follow from Eq. (11.28) and have the form

$$\ddot{q}_1 = 2e^{-q_1} - e^{-q_2}$$

$$\ddot{q}_2 = -e^{-q_1} + 2e^{-q_2} - e^{-q_3}$$

$$\ddot{q}_3 = -e^{-q_2} + 2e^{-q_3} - e^{-q_4}$$

.

.

.   (11.30)

.

$$\ddot{q}_{N-1} = -e^{-q_{N-2}} + 2e^{-q_{N-1}} - e^{-q_N}$$

$$\ddot{q}_N = -e^{-q_{N-1}} + 2e^{-q_N}$$

We recognize that this can be written compactly as

$$\ddot{q}_a = \sum_{b=1}^{N-1} K_{ab} e^{-q_b} \qquad (11.31)$$

where K_{ab} is the Cartan matrix for $SU(N)$ given in Eq. (11.17).

Let us note that a Lagrangian whose Euler-Lagrange variations gives rise to Eq. (11.31) is easily obtained to be

$$L = \sum_{a,b=1}^{N-1} \frac{1}{2} \dot{q}_a K_{ab}^{-1} \dot{q}_b - \sum_{a=1}^{N-1} e^{-q_a} \qquad (11.32)$$

where K_{ab}^{-1} is the inverse of the Cartan matrix. Let us note that the momenta conjugate to q_a are defined as

$$p_a = \frac{\partial L}{\partial \dot{q}_a} = \sum_{b=1}^{N-1} K_{ab}^{-1} \dot{q}_b \qquad (11.33)$$

Using the relations in Eq. (11.26), (11.29) and (10.9) we immediately see that

$$\{q_a, q_b\} = 0 = \{p_a, p_b\} \qquad (11.34)$$

and

$$\{q_a, p_b\} = \sum_{c=1}^{N-1} K_{bc}^{-1} \{q_a, \dot{q}_c\}$$

$$= \sum_{c=1}^{N-1} K_{bc}^{-1} \{Q_{a+1} - Q_a, P_{c+1} - P_c\}$$

$$= \sum_{c=1}^{N-1} K_{bc}^{-1}(2\delta_{ac} - \delta_{a,c+1} - \delta_{a+1,c})$$

or, $\quad \{q_a, p_b\} = \delta_{ab}$ $\qquad\qquad\qquad\qquad\qquad\qquad$ (11.35)

Here we have used the identity of Eq. (11.18) in the last step. Relations (11.34) and (11.35) show that (q_a, p_a) constitute a canonical coordinate system.

The Lax Pair:

So far we have introduced the group structure of SU(N) into the Toda system through the Cartan matrix. However, this seems quite accidental. To show that a deeper connection indeed exists, let us define the Lax operators

$$S = \frac{1}{2} \sum_{a=1}^{N-1} \left(p_a H_a + e^{-q_a/2}(E_a + E_{-a}) \right)$$

$$\qquad\qquad\qquad\qquad\qquad\qquad\qquad\qquad (11.36)$$

$$U = -\frac{1}{2} \sum_{a=1}^{N-1} e^{-q_a/2}(E_a - E_{-a})$$

Here H_a and E_a are the generators in the Chevalley basis defined in Eq. (11.24). Consequently, S and U are operators in the SU(N) algebra and have a $(N-1) \times (N-1)$ matrix form. Using the relations (11.25) we can show that

$$\frac{dS}{dt} - [S,U] = \frac{1}{2} \sum_{a=1}^{N-1} \left(\dot{p}_a H_a - \frac{1}{2} \dot{q}_a e^{-q_a/2}(E_a + E_{-a}) \right)$$

$$- \frac{1}{2} \left(\sum_{a=1}^{N-1} e^{-q_a} H_a - \frac{1}{2} \sum_{a,b=1}^{N-1} p_a e^{-q_b/2}(E_b + E_{-b}) K_{ba} \right)$$

Furthermore, using the definition of p_a from Eq. (11.33), we obtain

$$\frac{dS}{dt} - [S,U] = \frac{1}{2} \sum_{a,b=1}^{N-1} H_a K_{ab}^{-1} \left(\ddot{q}_b - \sum_{c=1}^{N-1} K_{bc} e^{-q_c} \right) \quad (11.37)$$

We recognize the term in the parenthesis on the right hand side as the equations (11.31). Consequently, the Lax equation

$$\frac{dS}{dt} = [S,U] \quad\quad\quad\quad\quad\quad (11.38)$$

indeed represents the Toda equations in the new variables. Following our discussion of Eq. (9.43) we recognize that the quantities

$$K_n = \frac{1}{n} \, \text{Tr} \, S^n \quad\quad\quad\quad\quad\quad (11.39)$$

must be conserved under the flow of the Toda equations. Furthermore, since the Lax operator S belongs to the $SU(N)$ algebra, the number of independent quantities of the form Eq. (11.39) can equal $N-1$ which is the rank of $SU(N)$. Let us recall that since we have already separated out the total momentum which is conserved, the number of independent conserved quantities really equals N which is the right number for integrability according to Liouville's criterion.

Before discussing the integrability of the Toda system, let us note that given the canonical Poisson brackets of

Eqs. (11.34) and (11.35) we can calculate

$$\{S_{ij}, S_{k\ell}\} \tag{11.40}$$

If we introduce a tensor notation

$$(A \otimes B)_{ik;j\ell} = A_{ij} B_{k\ell} \tag{11.41}$$

then it follows that

$$(A \otimes B)(C \otimes D) = AC \otimes BD \tag{11.42}$$

With this notation, all the components of the Poisson bracket in Eq. (11.40) can be calculated from Eqs. (11.34), (11.35) and (11.36) to be

$$\{S \underset{,}{\otimes} S\} = \frac{1}{4} \sum_{a,b=1}^{N-1} \left(H_a \otimes (E_b + E_{-b}) \{p_a , e^{-q_b/2}\} \right.$$

$$\left. + (E_a + E_{-a}) \otimes H_b \{e^{-q_a/2}, p_b\} \right)$$

or, $\{S \underset{,}{\otimes} S\}$

$$= \frac{1}{8} \sum_{a=1}^{N-1} e^{-q_a/2} \left(H_a \otimes (E_a + E_{-a}) - (E_a + E_{-a}) \otimes H_a \right) \tag{11.43}$$

This relation is quite useful in proving integrability as we will see later.

Integrability of the Toda system:

Since we have shown that the Toda system has the right number of linearly independent conserved quantities, integrability is proved if we can show that these conserved quantities are also in involution. To show this, let us introduce some more Lie algebraic structures. Let us define the following tensor product operators for SU(N)

$$C_+ = \sum_{\alpha>0} e_\alpha \otimes e_{-\alpha}$$

$$= \frac{1}{2} \sum_{\alpha>0} (\alpha,\alpha)\left(E_\alpha \otimes E_{-\alpha}\right) \qquad (11.44)$$

and

$$C_- = \sum_{\alpha<0} e_\alpha \otimes e_{-\alpha}$$

$$= \frac{1}{2} \sum_{\alpha<0} (\alpha,\alpha)\left(E_\alpha \otimes E_{-\alpha}\right) = \frac{1}{2} \sum_{\alpha>0} (\alpha,\alpha)\left(E_{-\alpha} \otimes E_\alpha\right) \quad (11.45)$$

Here the final expressions involve sums over only the positive roots of SU(N). Now, using Eqs. (11.23), (11.25) and (11.42), we see that

$$[C_+ , E_a \otimes I + I \otimes E_a]$$

$$= \sum_{\alpha>0} \frac{1}{2} (\alpha,\alpha)\left([E_\alpha,E_a] \otimes E_{-\alpha} + E_\alpha \otimes [E_{-\alpha},E_a]\right)$$

$$= -\frac{1}{2}(\alpha^a,\alpha^a)E_a \otimes H_a$$

$$+ \sum_{\alpha>0} \sqrt{\frac{(\alpha,\alpha)(\alpha+\alpha^a,\alpha+\alpha^a)}{2(\alpha^a,\alpha^a)}} \left(E_{\alpha+\alpha^a} \otimes E_{-\alpha}\right)$$

$$\left(N(\alpha,\alpha^a) + N(-\alpha-\alpha^a,\alpha^a)\right)$$

The second term on the right hand side vanishes upon using the identity in Eq. (11.13) so that

$$[C_+ , E_a \otimes I + I \otimes E_a] = -\frac{1}{2}(\alpha^a,\alpha^a)\left(E_a \otimes H_a\right) \quad (11.46)$$

Let us note here that in the above expression I stands for the identity operator. It can also be shown similarly that

$$[C_- , E_a \otimes I + I \otimes E_a] = -\frac{1}{2}(\alpha^a,\alpha^a)\left(H_a \otimes E_a\right) \quad (11.47)$$

If we now define

$$r = \frac{1}{2}(C_+ - C_-) \qquad\qquad (11.48)$$

and assume that the simple roots are normalized, namely,

$$(\alpha^a,\alpha^a) = 1 \qquad\qquad a = 1,2\ldots N-1 \quad (11.49)$$

then Eqs. (11.46) and (11.47) imply

$$[r\ ,\ E_a \otimes I + I \otimes E_a] = \tfrac{1}{4}\ (H_a \otimes E_a - E_a \otimes H_a) \qquad (11.50)$$

Similarly, it can be shown that

$$[r\ ,\ E_{-a} \otimes I + I \otimes E_{-a}] = \tfrac{1}{4}\ (H_a \otimes E_{-a} - E_{-a} \otimes H_a) \quad (11.51)$$

and

$$[r\ ,\ H_a \otimes I + I \otimes H_a] = 0 \qquad\qquad\qquad (11.52)$$

We are now ready to prove the integrability of the Toda system. Let us note from Eqs. (11.36) and (11.50)-(11.52) that

$$[r\ ,\ S \otimes I + I \otimes S]$$

$$= \tfrac{1}{2} \sum_{a=1}^{N-1} \left(p_a [r\ ,\ H_a \otimes I + I \otimes H_a] \right.$$

$$\left. + e^{-q_a/2} [r\ ,\ (E_a + E_{-a}) \otimes I + I \otimes (E_a + E_{-a})] \right)$$

$$= \tfrac{1}{8} \sum_{a=1}^{N-1} e^{-q_a/2} \left(H_a \otimes (E_a + E_{-a}) - (E_a + E_{-a}) \otimes H_a \right) \quad (11.53)$$

Comparing with Eq. (11.43), we see that

$$\{ S \underset{,}{\otimes} S \} = [r\ ,\ S \otimes I + I \otimes S] \qquad\qquad (11.54)$$

This is the fundamental relation in proving integrability just as the recursion relations of Eq. (9.49) were crucial in the geometric approach. It also emphasizes the interplay of the Hamiltonian methods with the group structure in that the left hand side only involves canonical Poisson bracket relations whereas the right hand side is determined from the group structure.

Note here that given Eq. (11.54) we can prove involution of the conserved quantities in the following way. First note that

$$\{\mathrm{Tr}\ s^n,\ s\}$$

$$= n\ \mathrm{Tr}_L\left((s^{n-1} \otimes I)\{s \underset{,}{\otimes} s\}\right) \tag{11.55}$$

where Tr_L denotes trace over the first element of the tensor product space. Using Eqs. (11.42) and (11.54), we see that

$$\{\mathrm{Tr}s^n,\ s\} = n\ \mathrm{Tr}_L\left((s^{n-1} \otimes I)[r\ ,\ s \otimes I + I \otimes s]\right)$$

$$= n\ \mathrm{Tr}_L\left((s^{n-1} \otimes I)[r\ ,\ I \otimes s]\right)$$

$$= n\left[\mathrm{Tr}_L s^{n-1} r\ ,\ s\right] \tag{11.56}$$

In deriving this we have used the cyclicity property of trace and an obvious notation

$$\mathrm{Tr}_L\ s^{n-1}\ r = \mathrm{Tr}_L(s^{n-1} \otimes I)r \tag{11.57}$$

It is obvious from Eq. (11.56) that

$$\{ \text{Tr } S^n, \text{ Tr } S^m \}$$

$$= m \text{ Tr } S^{m-1} \{ \text{Tr } S^n, S \}$$

$$= mn \text{ Tr}[\text{Tr}_L S^{n-1} r , S] = 0 \qquad \forall \; m,n \qquad (11.58)$$

The right hand side vanishes because of the cyclicity properties of trace.

This, therefore, shows that the conserved quantities constructed in Eq. (11.39) are also in involution and consequently, the Toda system is integrable. Note that each of the Hamiltonians generates its own flow and that the higher order Lax equations can be obtained from Eq. (11.56) as

$$\frac{dS}{dt_n} = \{ S, K_n \}$$

$$= \frac{1}{n} \{ S, \text{Tr } S^n \}$$

$$= [S, \text{Tr}_L S^{n-1} r]$$

$$\text{or, } \quad \frac{dS}{dt_n} = [S, \text{ Tr}_L S^{n-1} r] = [S, U_n] \qquad (11.59)$$

where

$$U_n = \text{Tr}_L S^{n-1} r = \text{Tr}_L (S^{n-1} \otimes I) r \qquad (11.60)$$

It is clear from our discussion that although the Toda lattice that we have described corresponds to the group SU(N), one could have in fact chosen any group including affine and Kac-Moody ones and the corresponding Toda lattices would also be integrable.

References:

Cahn, R., Semi-Simple Lie Algebras and their Representations, Benjamin/Cummings, 1984.

Faddeev, L. D., Soviet Sci. Reviews, C1, Ed. S. P. Novikov, 1981.

Faddeev, L. D. in Recent Advances in Field Theory and Statistical Mechanics, Les Houches Summer School Proc., Vol. 39, North Holland, 1984.

Georgi, H., Lie Algebras in Particle Physics, Benjamin/Cummings, 1982.

Humphreys, J. E., Introduction to Lie Algebras and Representation Theory, Springer-Verlag, 1972.

Kostant, B. in Lect. Notes Math. Vol. 170, Springer-Verlag, 1970.

Kostant, B., Adv. Math. 34, 195 (1979).

Leznov, A. M. and M. V. Savelier, Lett. Math. Phys. 3, 489 (1979); Comm. Math. Phys. 79, 11 (1980).

Olive, D. I. and N. Turok, Nucl. Phys. B220, 491 (1983).

CHAPTER 12

ZAKHAROV-SHABAT FORMULATION

So far, we have studied only two integrable models, namely, the continuum model of KdV and the finite dimensional Toda lattice. However, there exist several other models - at least continuum ones - which are also integrable. A natural question that arises, therefore, is whether a unified description of all such models is possible and surprisingly enough the answer is yes. In fact, in trying to understand the nonlinear Schrödinger equation, which is integrable, Zakharov and Shabat used an approach which was later generalized by Ablowitz-Kaup-Newell-Segur (AKNS) to describe various other integrable models. This approach uses a Lax operator which is first order in the derivative $\frac{\partial}{\partial x}$, in contrast to the second order formulation which we have already studied in Chapter 6. Besides describing various integrable models in a unified manner, this approach has the advantage that the inverse scattering method for such a formulation generalizes readily to the quantum case. In this chapter, therefore, we will describe the first order formulation of the Lax operator and study the inverse scattering method for the specific example of the nonlinear Schrödinger equation, in order to bring out the essential features of this approach.

The First Order Formulation:

In the first order formulation, the nonlinear evolution

equation is obtained as the compatibility condition or the
consistency condition for a pair of equations. To understand the motivation for this, let us recall our discussion
of the Lax method in Chapter 6 (See Eqs. (6.23)-(6.26)). We
know that if

$$L(t)\psi(t) = -\lambda\psi(t) \tag{12.1}$$

and

$$\frac{\partial L(t)}{\partial t} = [B(t),L(t)] \tag{12.2}$$

where

$$\frac{\partial \psi(t)}{\partial t} = B(t)\psi(t) \tag{12.3}$$

then the spectral parameter is independent of t, namely,

$$\frac{\partial \lambda}{\partial t} = 0 \tag{12.4}$$

We can, of course, check this very easily. For example,
taking the time derivative of Eq. (12.1), we obtain

$$\frac{\partial L(t)}{\partial t} \psi(t) + L(t) \frac{\partial \psi(t)}{\partial t} = - \frac{\partial \lambda}{\partial t} \psi(t) - \lambda \frac{\partial \psi(t)}{\partial t}$$

Using Eq. (12.3), we can write this as

$$- \frac{\partial \lambda}{\partial t} \, \psi(t) = \Big(\frac{\partial L(t)}{\partial t} + L(t)B(t) + \lambda B(t) \Big) \psi(t)$$

$$= \Big(\frac{\partial L(t)}{\partial t} + [L(t),B(t)] + B(t)(L(t)+\lambda) \Big) \psi(t)$$

$$= 0 \qquad\qquad\qquad\qquad (12.5)$$

$$\text{or,} \quad \frac{\partial \lambda}{\partial t} = 0 \qquad\qquad\qquad\qquad (12.6)$$

In obtaining Eq. (12.5), we have used Eqs. (12.1) and (12.2) in the final step.

This analysis is quite interesting for it allows us to turn the argument around and identify the Lax pair in the following way. Namely, let us assume that

$$L(t)\psi(t) = -\lambda \psi(t)$$

and $$(12.7)$$

$$\frac{\partial \psi(t)}{\partial t} = B(t)\psi(t)$$

with

$$\frac{\partial \lambda}{\partial t} = 0 \qquad\qquad\qquad\qquad (12.8)$$

If the compatibility condition for the two equations in Eq. (12.7) gives the nonlinear evolution equation under study, then L(t) and B(t) can be identified as the Lax pair of the system. This is obvious from our previous analysis because as we can see from Eq. (12.5), the compatibility condition in this case is precisely the Lax equation of Eq. (12.2).

This alternate formulation of the Lax equation as a compatibility condition is quite useful and is at the heart of the first order formulation.

In the first order formulation, we would like L(t) to be linear in $\frac{\partial}{\partial x}$. We recall from our experience with the Klein-Gordon and the Dirac equations that a second order scalar differential equation is often expressible as a first order matrix differential equation. Keeping this in mind let us generalize the two equations of Eq. (12.7) to (2×2) matrix equations. (We can use higher dimensional matrices, but this is the simplest and the most useful.) Let us recall that the Pauli matrices are defined as

$$\sigma_1 = \begin{pmatrix} 0 & 1 \\ 1 & 0 \end{pmatrix} \qquad \sigma_2 = \begin{pmatrix} 0 & -i \\ i & 0 \end{pmatrix} \qquad \sigma_3 = \begin{pmatrix} 1 & 0 \\ 0 & -1 \end{pmatrix} \qquad (12.9)$$

so that

$$\sigma_+ = \frac{1}{2} (\sigma_1 + i\sigma_2) = \begin{pmatrix} 0 & 1 \\ 0 & 0 \end{pmatrix}$$

$$\sigma_- = \frac{1}{2} (\sigma_1 - i\sigma_2) = \begin{pmatrix} 0 & 0 \\ 1 & 0 \end{pmatrix} \qquad (12.10)$$

It also follows from their explicit forms that

$$\sigma_+^2 = \sigma_-^2 = 0 \qquad\qquad \sigma_3^2 = I$$

$$\sigma_\pm \sigma_3 = \mp \sigma_\pm = -\sigma_3 \sigma_\pm \qquad\qquad (12.11)$$

$$\sigma_\pm \sigma_\mp = \frac{1}{2} (I \pm \sigma_3)$$

where I is the two dimensional identity matrix.

Defining now a two component column vector

$$\phi = \begin{pmatrix} \phi_1 \\ \phi_2 \end{pmatrix} \qquad\qquad (12.12)$$

we generalize the two equations of Eq. (12.7) to first order matrix equations as

$$\left(\sigma_3 \frac{\partial}{\partial x} - q\sigma_+ + r\sigma_-\right)\phi = -i\varsigma\phi$$

or, $\dfrac{\partial\phi}{\partial x} = (q\sigma_+ + r\sigma_- - i\varsigma\sigma_3)\phi$ $\qquad\qquad (12.13)$

and

$$\frac{\partial\phi}{\partial t} = (P\sigma_+ + Q\sigma_- + R\sigma_3)\phi \qquad\qquad (12.14)$$

The dynamical variables $q(x,t)$ and $r(x,t)$ do not depend on the spectral parameter ς which is assumed to be independent of x and t. The coefficient functions P, Q and R, on the other hand, do depend on ς and are functionals of $q(x,t)$ and $r(x,t)$.

The compatibility condition for Eqs. (12.13) and (12.14) can now be easily obtained. We note from Eq. (12.13) that

$$\frac{\partial^2\phi}{\partial t\partial x} = (q_t\sigma_+ + r_t\sigma_-)\phi + (q\sigma_+ + r\sigma_- - i\varsigma\sigma_3)\frac{\partial\phi}{\partial t}$$

$$= (q_t\sigma_+ + r_t\sigma_-)\phi + (q\sigma_+ + r\sigma_- - i\varsigma\sigma_3)(P\sigma_+ + Q\sigma_- + R\sigma_3)\phi$$

$$\text{or,} \quad \frac{\partial^2 \phi}{\partial t \partial x} = \Big((q_t - qR - i\varsigma P)\sigma_+ + (r_t + rR + i\varsigma Q)\sigma_-$$

$$+ \; (\frac{qQ}{2} - \frac{rP}{2})\sigma_3 + (\frac{qQ}{2} + \frac{rP}{2} - i\varsigma R)\Big)\phi \qquad (12.15)$$

Similarly, from Eq. (12.14) we obtain

$$\frac{\partial^2 \phi}{\partial x \partial t} = (P_x \sigma_+ + Q_x \sigma_- + R_x \sigma_3)\phi + (P\sigma_+ + Q\sigma_- + R\sigma_3) \frac{\partial \phi}{\partial x}$$

$$= (P_x \sigma_+ + Q_x \sigma_- + R_x \sigma_3)\phi + (P\sigma_+ + Q\sigma_- + R\sigma_3)(q\sigma_+ + r\sigma_- - i\varsigma\sigma_3)\phi$$

$$\text{or,} \quad \frac{\partial^2 \phi}{\partial x \partial t} = \Big((P_x + i\varsigma P + qR)\sigma_+ + (Q_x - i\varsigma Q - rR)\sigma_-$$

$$+ \; (R_x + \frac{rP}{2} - \frac{qQ}{2})\sigma_3 + (\frac{qQ}{2} + \frac{rP}{2} - i\varsigma R)\Big)\phi \qquad (12.16)$$

Comparing Eqs. (12.15) and (12.16), we obtain the compatibility condition to be

$$R_x = qQ - rP \qquad\qquad\qquad (12.17)$$

$$r_t = Q_x - 2rR - 2i\varsigma Q \qquad\qquad (12.18)$$

$$q_t = P_x + 2qR + 2i\varsigma P \qquad\qquad (12.19)$$

Thus following our earlier analysis, we conclude that if for some choice of the variables, Eqs. (12.17)-(12.19) describe a given nonlinear evolution equation, then Eqs. (12.13) and (12.14) would define the Lax pair appropriate for such a system.   In fact, let us write down explicitly, the form of the Lax pair in such a case

$$L = \frac{\partial}{\partial x} - A(x,\varsigma) = \frac{\partial}{\partial x} - q\sigma_+ - r\sigma_- + i\varsigma\sigma_3$$

and   (12.20)

$$B = P\sigma_+ + Q\sigma_- + R\sigma_3$$

so that

$$\frac{\partial L}{\partial t} = [B,L] \qquad\qquad (12.21)$$

Let us note here that B is unique only up to an additive constant, since a constant commutes with everything. Let us note further that the eigenvalue equation in such a case, Eq. (12.13), is a first order differential equation, in contrast to the more conventional second order equation in Eq. (6.61).

The Nonlinear Equations:

We are now in a position to describe various nonlinear integrable models in a unified way.

i) KdV:

Let us choose r=6. Then Eq. (12.18) gives

$$R = \frac{1}{12} Q_x - \frac{i\varsigma}{6} Q \qquad\qquad (12.22)$$

We also obtain from Eq. (12.17) the relation

$$P = \frac{1}{6} qQ - \frac{1}{72} Q_{xx} + \frac{i\varsigma}{36} Q_x \qquad (12.23)$$

Using Eqs. (12.22) and (12.23) in Eq. (12.19), we obtain the evolution equation to be

$$q_t = -\frac{1}{72} Q_{xxx} + \frac{1}{3} qQ_x + \frac{1}{6} q_xQ - \frac{\varsigma^2}{18} Q_x \qquad (12.24)$$

Furthermore, if we choose $q = -\frac{1}{36} u(x,t)$ (compare Eqs. Eqs. (12.22)-(12.24) with Eqs. (7.53)-(7.55)), we obtain

$$u_t = \frac{1}{2} Q_{xxx} + \frac{1}{3} uQ_x + \frac{1}{6} u_xQ + 2\varsigma^2 Q_x$$

$$= \frac{1}{2} \left(D^3 + \frac{1}{3} (Du+uD)\right)Q + 2\varsigma^2 DQ \qquad (12.25)$$

This is nothing other than the generating equation for the KdV hierarchy as we have seen in Eqs. (6.71) and (7.57).

ii) MKdV:

Let us choose $r(x,t) = q(x,t) = -\frac{i}{\sqrt{6}} v(x,t)$. Then from Eq. (12.17) we obtain

$$(P-Q) = -i \sqrt{6} \left(\frac{R_x}{v}\right) \qquad (12.26)$$

The difference of Eqs. (12.18) and (12.19) gives

$$(P+Q) = \frac{\sqrt{6}}{2\varsigma} \left(\frac{R_x}{v}\right)_x + \frac{2}{\sqrt{6}\varsigma} vR$$

so that

$$(P+Q)_x = \frac{\sqrt{6}}{2\varsigma} \left(\frac{R_x}{v}\right)_{xx} + \frac{2}{\sqrt{6}\varsigma} (vR)_x \qquad (12.27)$$

Finally, the sum of Eqs. (12.18) and (12.19), upon using Eqs. (12.26) and (12.27), gives the evolution equation to be

$$v_t = \frac{3i}{2\varsigma} \left(\frac{R_x}{v}\right)_{xx} + \frac{i}{\varsigma} (vR)_x + 6i\varsigma \left(\frac{R_x}{v}\right) \qquad (12.28)$$

It is worth comparing Eqs. (12.26)-(12.28) with Eqs. (7.59), (7.63) and (7.64) respectively. It is clear then that if we let

$$R = -i\varsigma\left(-4\varsigma^2 + \frac{1}{3} v^2(x,t)\right) \qquad (12.29)$$

then the evolution equation in (12.28) takes the form

$$v_t = v_{xxx} + v^2 v_x$$

which is the MKdV equation.

iii) Nonlinear Schrödinger equation (NSE):

Let us choose $q(x,t) = \sqrt{\kappa}\psi^*$ and $r(x,t) = \sqrt{\kappa}\psi$ where κ is an arbitrary constant parameter which can be both positive or negative.   Consistency   of   the   set   of   equations (12.17)-(12.19) requires that

$$Q = (\text{sign } \kappa)P^* \tag{12.30}$$

Furthermore, if we choose

$$R = 2i\varsigma^2 + i\kappa\, \psi^*\psi \tag{12.31}$$

then we obtain from Eq. (12.17) (using Eq. (12.30)) that

$$Q = i\sqrt{\kappa}\, \psi_x - 2\varsigma\, \sqrt{\kappa}\, \psi \tag{12.32}$$

Using Eqs. (12.30)-(12.32), the evolution equations are obtained from Eqs. (12.18) and (12.19) to be

$$i\psi_t = - \psi_{xx} + 2\kappa|\psi|^2\psi \tag{12.33}$$

$$i\psi_t^* = \psi_{xx} - 2\kappa|\psi|^2\psi^* \tag{12.34}$$

This is the nonlinear Schrödinger equation where we have denoted

$$|\psi|^2 = \psi^*\psi$$

and the parameter κ measures the strength of the nonlinear interaction.

iv) Sine-Gordon equation:

Let us choose

$$r(x,t) = -q(x,t) = \tfrac{1}{2}\, \omega_x(x,t) \tag{12.35}$$

Furthermore, if we let

$$P = Q = \frac{i}{4\varsigma}\, \sin\omega \tag{12.36}$$

then Eq. (12.17) gives

$$R = \frac{i}{4\varsigma}\, \cos\omega \tag{12.37}$$

Using these, we obtain the evolution equation, from Eqs. (12.18) and (12.19), to be

$$\omega_{xt} = \sin\omega \tag{12.38}$$

This is the Sine-Gordon equation in the light cone variables as we have seen in Eq. (8.21).

v)   Sinh-Gordon equation:

We can similarly choose

$$r(x,t) = q(x,t) = \frac{1}{2}\, \omega_x(x,t) \qquad\qquad (12.39)$$

and let

$$Q = -P = \frac{i}{4\varsigma}\, \sinh\omega \qquad\qquad (12.40)$$

Then Eq. (12.17) determines the form of R to be

$$R = \frac{i}{4\varsigma}\, \cosh\omega \qquad\qquad (12.41)$$

Using these, we obtain the equation of motion from Eqs. (12.18) and (12.19), to be

$$\omega_{xt} = \sinh\omega \qquad\qquad (12.42)$$

which is the Sinh-Gordon equation in the light cone variables.

This discussion shows how various integrable models can be described in a unified manner in the first order formulation. Furthermore, it also brings out a very basic feature of the Lax equation, namely, whereas the Lax operators are functions of the spectral parameter, the consistency condition or the nonlinear evolution equation does not depend on it.

NSE as a Hamiltonian System:

In the rest of this chapter, we will study the specific example of the nonlinear Schrödinger equation to further elucidate the essential features of the first order formulation. First, let us show that the NS equation is a Hamiltonian system. From the form of the equations in Eqs. (12.33) and (12.34), we see that they can be obtained from the Euler-Lagrange variation of the following first order Lagrangian.

$$L = \frac{i}{2} \int_{-\infty}^{\infty} dx (\psi^* \psi_t - \psi_t^* \psi) - H \qquad (12.43)$$

where the Hamiltonian is given by

$$H = \int_{-\infty}^{\infty} dx (\psi_x^* \psi_x + \kappa (\psi^* \psi)^2) \qquad (12.44)$$

Here we are assuming that the dynamical variables $\psi(x,t)$ and $\psi^*(x,t)$ vanish asymptotically.

Following our discussion of Eqs. (9.26), (9.27) and (9.30), we can read off the symplectic structure and the Poisson bracket structure from the Lagrangian in Eq. (12.43). In fact, let us note that if

$$\{\psi(x,t),\psi^*(y,t)\} = -i\delta(x-y)$$

(12.45)

$$\{\psi(x,t),\psi(y,t)\} = 0 = \{\psi^*(x,t),\psi^*(y,t)\}$$

then

$$\psi_t = \{\psi(x,t),H\} = i\psi_{xx} - 2i\kappa|\psi|^2\psi$$

or, $i\psi_t = -\psi_{xx} + 2\kappa|\psi|^2\psi$

(12.46)

and similarly,

$$\psi_t^* = \{\psi^*(x,t),H\} = -i\psi_{xx}^* + 2i\kappa|\psi|^2\psi^*$$

or, $i\psi_t^* = \psi_{xx}^* - 2\kappa|\psi|^2\psi^*$

(12.47)

These are, of course, the nonlinear Schrödinger equations written in the Hamiltonian form. Consequently, we conclude that NSE is a Hamiltonian system.

Time Development of the Scattering Data:

Let us recall from our discussions in Chapters 4 and 5 that the initial value problem for a nonlinear integrable system is best studied by the method of inverse scattering. This, as we have seen, is because whereas the nonlinear evolution is complicated, the scattering data of the associated linear equation correspond to the action-angle variables of the nonlinear system and, therefore, have a

very simple time development.   In fact, let us note from
Eqs. (12.13), (12.14) and (12.30)-(12.32) that for $\kappa < 0$, the
associated first order equations are given by

$$\left(\frac{\partial}{\partial x} - i\sqrt{|\kappa|} \ (\psi^* \sigma_+ + \psi \sigma_-) + i\varsigma \sigma_3\right)\phi = 0 \qquad (12.48)$$

and

$$\frac{\partial \phi}{\partial t} = B\phi = (P\sigma_+ + Q\sigma_- + R\sigma_3)\phi \qquad (12.49)$$

with

$$P = \sqrt{|\kappa|} \ \psi_x^* - 2i\varsigma \ \sqrt{|\kappa|} \ \psi^*$$

$$Q = -\sqrt{|\kappa|} \ \psi_x - 2i\varsigma \ \sqrt{|\kappa|} \ \psi \qquad (12.50)$$

$$R = 2i\varsigma^2 - i|\kappa| \ \psi^* \psi$$

Since we require $\psi(x,t)$, $\psi^*(x,t)$ and their derivatives to
vanish asymptotically, it is clear from Eq. (12.50) that

$$P \xrightarrow[|x| \to \infty]{} 0$$

$$Q \xrightarrow[|x| \to \infty]{} 0 \qquad (12.51)$$

$$R \xrightarrow[|x| \to \infty]{} 2i\varsigma^2$$

We also recall that the matrix B is unique only up to an
additive constant.   Allowing for this, then, the time
evolution equation, Eq. (12.49), takes the asymptotic form

$$\frac{\partial \phi}{\partial t} = \left(C + 2i\varsigma^2 \sigma_3\right)\phi \qquad \text{as} \qquad |x| \to \infty \qquad (12.52)$$

where the constant C can be determined from the boundary
conditions of the eigenvalue problem.   The asymptotic
evolution is clearly simple and since the scattering data is
related to the asymptotic behavior of the wave function, we
see that the time development of the scattering data must
also be necessarily simple.

Let us next turn to the eigenvalue equation, Eq.
(12.48).   Let us note some of its properties before
analyzing the scattering data.   First of all, note that
unlike the Schrödinger equation, this is not Hermitian.
Consequently, the spectral parameter ς need not be real.
Secondly, if the two component vector ϕ (see Eq. (12.12)) is
a solution of Eq. (12.48), then it can be easily shown using
the properties of the Pauli matrices that

$$\left(\frac{\partial}{\partial x} - i\sqrt{|\kappa|} \, (\psi^* \sigma_+ + \psi \sigma_-) + i\varsigma^* \sigma_3\right)(i\sigma_2 \phi^*) = 0 \quad (12.53)$$

In other words, if $\phi(x,t)$ with the form

$$\phi = \begin{pmatrix} \phi_1 \\ \phi_2 \end{pmatrix} \qquad\qquad (12.54)$$

is a solution of Eq. (12.48) with the value of the spectral
parameter ς , then

$$\tilde{\phi} = i\sigma_2\phi^* = \begin{pmatrix} \phi_2^* \\ -\phi_1^* \end{pmatrix} \qquad (12.55)$$

is also a solution of Eq. (12.48) with the value of the
spectral parameter ς^*. It also follows from the definition
of $\tilde{\phi}$ and Eq. (12.52) that asymptotically $\tilde{\phi}$ satisfies the
evolution equation

$$\frac{\partial\tilde{\phi}}{\partial t} = \left(c^* + 2i\varsigma^2\sigma_3\right)\tilde{\phi} \qquad as \qquad |x| \to \infty \qquad (12.56)$$

Let us also note here that

$$\tilde{\tilde{\phi}} = i\sigma_2\tilde{\phi}^* = -\phi$$

We can derive the analogue of the Wronskian for the
present case in the following way. Let us assume that ϕ and
ω are two solutions of Eq. (12.48) with the spectral
parameters ς_1 and ς_2 respectively. That is,

$$\left(\frac{\partial}{\partial x} - i\sqrt{|\kappa|} \ (\psi^*\sigma_+ + \psi\sigma_-) + i\varsigma_1\sigma_3\right)\phi = 0 \qquad (12.57)$$

and

$$\left(\frac{\partial}{\partial x} - i\sqrt{|\kappa|} \ (\psi^*\sigma_+ + \psi\sigma_-) + i\varsigma_2\sigma_3\right)\omega = 0 \qquad (12.58)$$

Multiplying Eq. (12.57) with $i\omega^T\sigma_2$ and Eq. (12.58) with
$i\phi^T\sigma_2$ (T denotes transposition) and subtracting one from the
other, we can show using the properties of the Pauli

matrices that

$$\frac{\partial}{\partial x} \left(\omega^T i \sigma_2 \phi \right) - i (\varsigma_1 - \varsigma_2) \omega^T \sigma_1 \phi = 0 \qquad (12.59)$$

Written out explicitly, in terms of the components, this has the form

$$\frac{\partial}{\partial x} (\omega_1 \phi_2 - \omega_2 \phi_1) - i (\varsigma_1 - \varsigma_2)(\omega_1 \phi_2 + \omega_2 \phi_1) = 0 \qquad (12.60)$$

If we now assume that ϕ and ω correspond to the same value of the spectral parameter, namely $\varsigma_1 = \varsigma_2$, and define the analogue of the Wronskian as

$$W(\phi, \omega) = \phi^T i \sigma_2 \omega = \phi_1 \omega_2 - \phi_2 \omega_1 = -W(\omega, \phi) \qquad (12.61)$$

then it is clear from Eqs. (12.59) and (12.60) that $W(\phi, \omega)$ is a constant in space since

$$\frac{\partial W(\phi, \omega)}{\partial x} = 0 \qquad (12.62)$$

Let us next introduce the Jost functions for this system. Let f and g be the solutions of Eq. (12.48) for real values of the spectral parameter with the following asymptotic behavior.

$$f(x, \varsigma) \xrightarrow[x \to \infty]{} \binom{0}{1} e^{i \varsigma x}$$

$$(12.63)$$

$$g(x, \varsigma) \xrightarrow[x \to -\infty]{} \binom{-1}{0} e^{-i \varsigma x}$$

$\tilde{f}$ and $\tilde{g}$ can then be seen from Eq. (12.55) to take the form

$$\tilde{f}(x,\varsigma) \xrightarrow[x\to\infty]{} \begin{pmatrix} 1 \\ 0 \end{pmatrix} e^{-i\varsigma x}$$

$$\tilde{g}(x,\varsigma) \xrightarrow[x\to-\infty]{} \begin{pmatrix} 0 \\ 1 \end{pmatrix} e^{i\varsigma x}$$

(12.64)

Here we have used the fact that ς is chosen to be real for the definition of the Jost functions. Furthermore, since we require the Jost functions to retain their asymptotic behavior for all values of t, we can determine their asymptotic evolution, using Eqs. (12.52) and (12.56), to be (see also Eq. (13.32) in the next chapter)

$$\frac{\partial f}{\partial t} = 2i\varsigma^2(I+\sigma_3)f$$

$$\frac{\partial \tilde{f}}{\partial t} = -2i\varsigma^2(I-\sigma_3)\tilde{f}$$

(12.65)

$$\frac{\partial g}{\partial t} = -2i\varsigma^2(I-\sigma_3)g$$

$$\frac{\partial \tilde{g}}{\partial t} = 2i\varsigma^2(I+\sigma_3)\tilde{g}$$

The functions $f(x,\varsigma)$ and $\tilde{f}(x,\varsigma)$ form a complete set of solutions of Eq. (12.48) just as $g(x,\varsigma)$ and $\tilde{g}(x,\varsigma)$ do. Therefore, we can expand

$$f(x,\varsigma) = a(\varsigma)\tilde{g}(x,\varsigma) + b(\varsigma)g(x,\varsigma)$$

$$\tilde{f}(x,\varsigma) = -\tilde{a}(\varsigma)g(x,\varsigma) + \tilde{b}(\varsigma)\tilde{g}(x,\varsigma)$$

(12.66)

where the coefficient functions can be readily determined to be

$$a(\varsigma) = W(f(x,\varsigma),g(x,\varsigma)) = f_1 g_2 - f_2 g_1$$

$$b(\varsigma) = -W(f(x,\varsigma),\tilde{g}(x,\varsigma)) = f_2 \tilde{g}_1 - f_1 \tilde{g}_2$$

(12.67)

$$\tilde{a}(\varsigma) = W(\tilde{f}(x,\varsigma),\tilde{g}(x,\varsigma)) = \tilde{f}_1 \tilde{g}_2 - \tilde{f}_2 \tilde{g}_1$$

$$\tilde{b}(\varsigma) = W(\tilde{f}(x,\varsigma),g(x,\varsigma)) = \tilde{f}_1 g_2 - \tilde{f}_2 g_1$$

Let us note that, since by definition

$$\tilde{f}(x,\varsigma) = i\sigma_2 f^*(x,\varsigma)$$

$$\tilde{g}(x,\varsigma) = i\sigma_2 g^*(x,\varsigma)$$

it follows from Eq. (12.66) that for real ς

$$\tilde{a}(\varsigma) = a^*(\varsigma)$$

$$\tilde{b}(\varsigma) = b^*(\varsigma)$$

(12.68)

In general for complex values of the spectral parameter, it can be shown that

$$\tilde{a}(\varsigma) = a^*(\varsigma^*)$$

$$\tilde{b}(\varsigma) = b^*(\varsigma^*)$$

(12.69)

Let us also note from the expressions in Eq. (12.67) that

$$a(\varsigma)\tilde{a}(\varsigma) + b(\varsigma)\tilde{b}(\varsigma) = W(f,\tilde{f})W(g,\tilde{g}) = 1 \qquad (12.70)$$

Using Eq. (12.68), we can write this as

$$|a(\varsigma)|^2 + |b(\varsigma)|^2 = 1 \qquad\qquad (12.71)$$

This is the analogue of the unitarity relation in Eq. (5.12) for the present case.

Our discussion, in Chapter 5, on the analytic behavior of the coefficient functions, can now be generalized to the present case. In particular, it can be shown that $a(\varsigma)$ is analytic in the upper half of the complex ς-plane where $\text{Im}\varsigma > 0$ and that the discrete eigenvalues of Eq. (12.48) are obtained from

$$a(\varsigma_n) = 0 \qquad\qquad \text{Im } \varsigma_n > 0$$

where   (12.72)

$$f(x,\varsigma_n) = b_n g(x,\varsigma_n) \qquad\qquad n = 1,2,\ldots N$$

Let us note here, for future use, that the relations in Eq. (12.66) can be written in the matrix form as

$$
\begin{pmatrix} \tilde{f}_1 & f_1 \\ \tilde{f}_2 & f_2 \end{pmatrix} = \begin{pmatrix} -g_1 & \tilde{g}_1 \\ -g_2 & \tilde{g}_2 \end{pmatrix} \begin{pmatrix} a^*(\varsigma) & -b(\varsigma) \\ b^*(\varsigma) & a(\varsigma) \end{pmatrix}
\qquad (12.73)
$$

Let us define

$$
F = \begin{pmatrix} \tilde{f}_1 & f_1 \\ \tilde{f}_2 & f_2 \end{pmatrix} \quad \text{and} \quad G = \begin{pmatrix} -g_1 & \tilde{g}_1 \\ -g_2 & \tilde{g}_2 \end{pmatrix}
\qquad (12.74)
$$

such that

$$
\lim_{x \to \infty} F(x,\varsigma) = \lim_{x \to -\infty} G(x,\varsigma) \to \begin{pmatrix} e^{-i\varsigma x} & 0 \\ 0 & e^{i\varsigma x} \end{pmatrix} = e^{-i\varsigma\sigma_3 x}
\qquad (12.75)
$$

Furthermore, let us define

$$
T(\varsigma) = \begin{pmatrix} a(\varsigma) & b(\varsigma) \\ -b^*(\varsigma) & a^*(\varsigma) \end{pmatrix}
\qquad (12.76)
$$

It can be readily checked, using Eq. (12.71), that the determinant of this matrix is unity so that the inverse exists and has the form

$$
T^{-1}(\varsigma) = \begin{pmatrix} a^*(\varsigma) & -b(\varsigma) \\ b^*(\varsigma) & a(\varsigma) \end{pmatrix}
\qquad (12.77)
$$

In terms of these matrices, Eq. (12.73) can be written as

$$F(x,\varsigma) = G(x,\varsigma)T^{-1}(\varsigma) \qquad (12.78)$$

Conversely,

$$G(x,\varsigma) = F(x,\varsigma)T(\varsigma) \qquad (12.79)$$

Written out explicitly, this gives

$$g(x,\varsigma) = -a(\varsigma)\tilde{f}(x,\varsigma) + b^*(\varsigma)f(x,\varsigma)$$

$$\tilde{g}(x,\varsigma) = a^*(\varsigma)f(x,\varsigma) + b(\varsigma)\tilde{f}(x,\varsigma)$$

$$(12.80)$$

Let us note here that both $F(x,\varsigma)$ and $G(x,\varsigma)$ can be thought of as matrix solutions of Eq. (12.48) since each column vector is a solution and $T(\varsigma)$ is known as the transition matrix.

We are now in a position to calculate the time evolution of the scattering data.   Let ·us note from Eq. (12.67) and the asymptotic evolution of the Jost functions in Eq. (12.65) that

$$\frac{\partial a(\varsigma,t)}{\partial t} = \lim_{x \to \infty} \frac{\partial}{\partial t} (f_1 g_2 - f_2 g_1)$$

$$= \lim_{x \to \infty} -\frac{\partial}{\partial t} (f_2 g_1) = 0 \qquad (12.81)$$

and

$$\frac{\partial b(\varsigma,t)}{\partial t} = \underset{x \to \infty}{Lim} \ \frac{\partial}{\partial t} \ (f_2 \tilde{g}_1 - f_1 \tilde{g}_2)$$

$$= \underset{x \to \infty}{Lim} \ \frac{\partial}{\partial t} \ (f_2 \tilde{g}_1)$$

$$= \underset{x \to \infty}{Lim} \ 4i\varsigma^2 f_2 g_1 = 4i\varsigma^2 b(\varsigma,t) \qquad (12.82)$$

Since $b_n(t)$ is the analytic continuation of $b(\varsigma,t)$ to $\varsigma = \varsigma_n$, we conclude that

$$\frac{\partial b_n(t)}{\partial t} = 4i\varsigma_n^2 b_n(t) \qquad (12.83)$$

Furthermore, we recall from Eq. (5.32) that b_n is directly related to the normalization of the bound state wave functions and consequently, Eqs. (12.81)-(12.83) completely determine the time evolution of the scattering data, namely,

$$a(\varsigma,t) = a(\varsigma,0)$$

$$b(\varsigma,t) = b(\varsigma,0)e^{4i\varsigma^2 t} \qquad (12.84)$$

$$b_n(t) = b_n(0)e^{4i\varsigma_n^2 t}$$

Given this, one can use the Gel'fand-Levitan-Marchenko equation to determine the time evolution of $\psi(x,t)$ and $\psi^*(x,t)$. The only difference from the methods of Chapter 4 and 5 here is that the integral equation in the present case would involve matrices.

References:

Ablowitz, M. J., D. J. Kaup, A. C. Newell and H. Segur, Phys. Rev. Lett. 30, 1262 (1973); Phys. Rev. Lett. 31, 125 (9173).

Ablowitz, M. J., D. J. Kaup, A. C. Newell and H. Segur, Stud. Appl. Math. 53, 249 (1974).

Faddeev, L. D., Sov. Sci. Reviews C1, Ed. S. P. Novikov, 1981.

Faddeev, L. D. in Recent Advances in field Theory and Statistical Mechanics, Les Houches Summer School Proc., Vol. 39, North Holland 1984.

Faddeev, L. D. and L. A. Takhtajan, Hamiltonian Methods in the Theory of Solitons, Springer-Verlag, 1987.

Flaschka, H. and A. C. Newell in Dynamical Systems, Theory and Applications, Ed. J. Moser, Springer-Verlag, 1974.

Novikov, S., S. V. Manakov, L. P. Pitaevskii and V. E. Zakharov, Theory of solitons, Consultants Bureau, 1984.

Zakharov, V. E. and A. B. Shabat, JETP 34, 62 (1972).

Zakharov, V. E. and S. V. Manakov, Theo. Math. Phys. 19, 551 (1974).

Zakharov, V. E., L. A. Takhtadzhyan and L. D. Faddeev, Sov. Phys. Dokl. 19, 824 (1975).

CHAPTER 13

THE ZERO CURVATURE METHOD

In the last chapter, we discussed the generalized Zakharov-Shabat or the AKNS system of equations. We studied the time development of the scattering data specifically for the nonlinear Schrödinger equation when the interaction parameter, κ, is negative. Following our discussion in the KdV case (see the appendix), we can calculate explicitly the Poisson brackets between $a(\varsigma)$ and $b(\varsigma)$ and consequently identify the action-angle variables to prove the integrability of the system. However, the Poisson brackets can be calculated in a more elegant way, using the zero curvature condition. In this chapter, we will develop the method of zero curvature in detail since it generalizes readily to the corresponding quantum systems. Let us note here that although our discussion would be primarily within the context of the nonlinear Schrödinger equation, the method is quite general and applies to any system that can be described by the Zakharov-Shabat formulation.

The Zero Curvature Condition:

We have already discussed, in some detail, the zero curvature condition in Chapter 7 (see Eq. (7.49)). Let us observe here that the Lax equation and the zero curvature condition are synonymous in the first order formulation. For example, let us note that the Lax equation

$$\frac{\partial L}{\partial t} = [B, L] \qquad (13.1)$$

can also be written as

$$\left[\frac{\partial}{\partial t} - B, L\right] = 0 \tag{13.2}$$

Recalling the form of L from Eq. (12.20) for the AKNS system, this equation becomes

$$\left[\frac{\partial}{\partial t} - B(x,\varsigma) \ , \ \frac{\partial}{\partial x} - A(x,\varsigma)\right] = 0$$

or, $$\frac{\partial A(x,\varsigma)}{\partial t} - \frac{\partial B(x,\varsigma)}{\partial x} - [B(x,\varsigma),A(x,\varsigma)] = 0 \tag{13.3}$$

Let us note that this is the zero curvature condition of Eq. (7.49), namely,

$$F_{\mu\nu} = \partial_\mu A_\nu - \partial_\nu A_\mu + [A_\mu, A_\nu] = 0 \qquad \mu,\nu = 0,1 \tag{13.4}$$

if we identify

$$A_0 = -B(x,\varsigma)$$
$$\tag{13.5}$$
$$A_1 = -A(x,\varsigma)$$

The consequences of zero curvature are far reaching. For example, let us define the path ordered exponential

$$U_\gamma(x_2,t_2;x_1,t_1) = P_\gamma e^{-\int_{(x_1,t_1)}^{(x_2,t_2)} dx^\mu A_\mu} \qquad (13.6)$$

where P_γ stands for ordering of the points along the path of integration, γ, such that points closer to (x_2,t_2) stand to the left of those closer to (x_1,t_1).   If we now take the product of two such exponentials, namely,

$$U_{\gamma_1}(x_2,t_2;x_1,t_1)\ U_{\gamma_2}(x_1,t_1;x_2,t_2)$$

then it is straightforward to see using the Baker-Cambell-Hausdorf formula and the Stoke's theorem that

$$U_{\gamma_1}(x_2,t_2;x_1,t_1)\ U_{\gamma_2}(x_1,t_1;x_2,t_2) = e^{-\frac{1}{2}\int_C d\sigma^{\mu\nu} F_{\mu\nu}} \qquad (13.7)$$

where the integration on the right is over the area C enclosed by the closed path $\gamma_1+\gamma_2$. It is clear, therefore, that if the curvature vanishes, namely if Eq. (13.4) is satisfied, then

$$U_{\gamma_1}(x_2,t_2;x_1,t_1)\ U_{\gamma_2}(x_1,t_1;x_2,t_2) = I \qquad (13.8)$$

Noting from Eq. (13.6) that

$$U_\gamma^{-1}(x_2,t_2;x_1,t_1) = U_\gamma(x_1,t_1;x_2,t_2) \qquad (13.9)$$

Eq. (13.8) then leads to

$$U_{\gamma_1}\left(x_2, t_2; x_1, t_1\right) = U_{\gamma_2}\left(x_2, t_2; x_1, t_1\right) \qquad (13.10)$$

In other words, if the zero curvature condition is satisfied, then the path ordered exponential is independent of the path and depends only on the end points. Consequently, it follows that for a closed path

$$U(x, t; x, t) = I \qquad (13.11)$$

In what follows, we will suppress the path ordering symbol for convenience.

The Transition Matrix:

An object of fundamental importance in the study of integrable systems is given by

$$T(x, y; \varsigma) = U(x, t; y, t) = e^{-\int_y^x dz A_1(z)}$$

$$\text{or,} \quad T(x, y; \varsigma) = e^{\int_y^x dz A(z, \varsigma)} \qquad (13.12)$$

In writing the last line, we have used the identification

given in Eq. (13.5). To understand the role of $T(x,y;\varsigma)$, let us analyze the linear eigenvalue problem in Eq. (12.13) (also known as the auxiliary linear problem), namely,

$$\frac{\partial \phi(x)}{\partial x} = A(x,\varsigma)\phi(x) \qquad\qquad (13.13)$$

As can be directly checked, the formal solution of Eq. (13.13) is given by

$$\phi(x) = T(x,y;\varsigma)\phi(y) \qquad\qquad (13.14)$$

That is, $T(x,y;\varsigma)$ translates the solutions of the auxiliary problem along the x-axis for a fixed time. We see from the definition in Eq. (13.12) as well as from Eqs. (13.9) and (13.11) that

$$T(x,y;\varsigma)T(y,z;\varsigma) = T(x,z;\varsigma)$$

$$T^{-1}(x,y;\varsigma) = T(y,x;\varsigma) \qquad\qquad (13.15)$$

$$T(x,x;\varsigma) = I$$

Furthermore, we note from Eq. (13.12) that $T(x,y;\varsigma)$ also satisfies the auxiliary problem, namely,

$$\frac{\partial T(x,y;\varsigma)}{\partial x} = A(x,\varsigma)T(x,y;\varsigma) \qquad\qquad (13.16)$$

It, then, follows from Eq. (13.15) that

$$\frac{\partial T(x,y;\varsigma)}{\partial y} = -T(x,y;\varsigma)A(y,\varsigma) \qquad\qquad (13.17)$$

Let us recall that $A(x,\varsigma)$ is a matrix. For example, for the nonlinear Schrödinger equation with $\kappa < 0$, its form is (see Eq. (12.48))

$$A(x,\varsigma) = i\sqrt{|\kappa|} \ (\psi^*\sigma_+ + \psi\sigma_-) - i\varsigma\sigma_3 \tag{13.18}$$

Consequently, $T(x,y;\varsigma)$ is in general a matrix and from Eqs. (13.16) and (13.17) we see that it is a matrix solution of the auxiliary problem. Let us note that the unperturbed solution of Eqs. (13.16) and (13.17) (namely, with $\psi(x) = \psi^*(x) = 0$) is given by

$$T_0(x,y;\varsigma) = T_0(x-y;\varsigma) = e^{-i\varsigma\sigma_3(x-y)} \tag{13.19}$$

On the other hand, since $\psi(x)$ and $\psi^*(x)$ vanish asymptotically, we also see from the definitions in Eqs. (13.12) and (13.18) that

$$\lim_{x,y\to\infty} T(x,y;\varsigma) = e^{-i\varsigma\sigma_3(x-y)}$$

and $\tag{13.20}$

$$\lim_{x,y\to-\infty} T(x,y;\varsigma) = e^{-i\varsigma\sigma_3(x-y)}$$

Let us next define the matrix functions

$$F(x,\varsigma) = \operatorname*{Lim}_{y \to \infty} T(x,y;\varsigma)T_0(y;\varsigma)$$

and   (13.21)

$$G(x,\varsigma) = \operatorname*{Lim}_{y \to -\infty} T(x,y;\varsigma)T_0(y;\varsigma)$$

It is straightforward to see that $F(x,\varsigma)$ and $G(x,\varsigma)$ satisfy Eq. (13.13) with the following asymptotic behavior.

$$\operatorname*{Lim}_{x \to \infty} F(x,\varsigma) = \operatorname*{Lim}_{x,y \to \infty} T(x,y;\varsigma)T_0(y;\varsigma) \longrightarrow e^{-i\varsigma\sigma_3 x} \quad (13.22)$$

and

$$\operatorname*{Lim}_{x \to -\infty} G(x,\varsigma) = \operatorname*{Lim}_{x,y \to -\infty} T(x,y;\varsigma)T_0(y;\varsigma) \longrightarrow e^{-i\varsigma\sigma_3 x} \quad (13.23)$$

We recognize, from Eq. (12.75), that these are precisely the asymptotic relations satisfied by the matrix Jost solutions of Eq. (13.13) already defined in Eq. (12.74). Consequently, we identify $F(x,\varsigma)$ and $G(x,\varsigma)$ as the appropriate matrix Jost solutions of the auxiliary problem.   We recall from Eq. (12.79) that the transition matrix, $T(\varsigma)$, is defined by

$$G(x,\varsigma) = F(x,\varsigma)T(\varsigma)$$

$$\text{or,} \quad T(\varsigma) = F^{-1}(x,\varsigma)G(x,\varsigma) \qquad\qquad (13.24)$$

Substituting the forms of $F(x,\varsigma)$ and $G(x,\varsigma)$ from Eq. (13.21) and using the properties of $T(x,y;\varsigma)$ given in Eq. (13.15), we obtain

$$T(\varsigma) = \lim_{\substack{y\to\infty \\ z\to-\infty}} T_0^{-1}(y;\varsigma)T^{-1}(x,y;\varsigma)T(x,z;\varsigma)T_0(z;\varsigma)$$

$$= \lim_{\substack{y\to\infty \\ z\to-\infty}} T_0^{-1}(y;\varsigma)T(y,x;\varsigma)T(x,z;\varsigma)T_0(z;\varsigma)$$

$$\text{or,} \quad T(\varsigma) = \lim_{\substack{y\to\infty \\ z\to-\infty}} e^{i\varsigma\sigma_3 y}T(y,z;\varsigma)e^{-i\varsigma\sigma_3 z} \tag{13.25}$$

This is also defined in Eq. (12.76) in terms of the scattering data as

$$T(\varsigma) = \begin{pmatrix} a(\varsigma) & b(\varsigma) \\ -b^*(\varsigma) & a^*(\varsigma) \end{pmatrix} \tag{13.26}$$

We note here for completeness that most of the models we have discussed have been defined on the real line. This is because the periodic problems require special care. The formulation of this chapter, however, readily extends to the periodic systems where the transition matrix becomes related to the monodromy matrix for the system of equations.

Time Evolution of the Transition Matrix:

We have seen earlier, in Eq. (12.49), that the solutions

of Eq. (13.13) satisfy the time evolution equation

$$\frac{\partial \phi(x)}{\partial t} = B(x,\varsigma)\phi(x) + C\phi(x) \qquad (13.27)$$

where C is a constant.   We expect the matrix solutions of
Eq. (13.13) to   satisfy   a   similar   equation.   Thus
generalizing C to a matrix function of y we write

$$\frac{\partial T(x,y;\varsigma)}{\partial t} = B(x,\varsigma)T(x,y;\varsigma) + T(x,y;\varsigma)C(y) \qquad (13.28)$$

Consistency of the relations   in   Eq. (13.15) determines the
matrix function to be

$$C(y) = -B(y,\varsigma) \qquad (13.29)$$

so that the   equation   for   the   time   evolution of $T(x,y;\varsigma)$
takes the form

$$\frac{\partial T(x,y;\varsigma)}{\partial t} = B(x,\varsigma)T(x,y;\varsigma) - T(x,y;\varsigma)B(y,\varsigma) \qquad (13.30)$$

Let us recall from Eqs. (12.49)-(12.51) that asymptotically

$$\underset{|x| \to \infty}{\text{Lim}} B(x,\varsigma) = 2i\varsigma^2\sigma_3 \qquad (13.31)$$

so that we have the asymptotic relation

$$\underset{|x|,|y|\to\infty}{\text{Lim}} \frac{\partial T(x,y;\varsigma)}{\partial t} = 2i\varsigma^2[\sigma_3,T(x,y;\varsigma)] \qquad (13.32)$$

(Compare this with Eq. (12.65).)   The time evolution of the transition matrix is now easily obtained from Eqs. (13.25) and (13.31) to be

$$\frac{\partial T(\varsigma)}{\partial t} = 2i\varsigma^2[\sigma_3,T(\varsigma)] \qquad (13.33)$$

Using the representation in Eq. (13.26) for the transition matrix, we then obtain

$$\frac{\partial a(\varsigma,t)}{\partial t} = 0$$

$$\qquad (13.34)$$

$$\frac{\partial b(\varsigma,t)}{\partial t} = 4i\varsigma^2 b(\varsigma,t)$$

which is the same as the earlier result of Eqs. (12.81) and (12.82).

   The time evolution of the transition matrix, Eq. (13.33), is in the form of a Lax equation. Consequently, following our discussion in Eqs. (9.43)-(9.45), we see that all quantities of the form

$$K_n = \frac{1}{n} \text{Tr} \ (T(\varsigma))^n \qquad\qquad n = \pm 1, \pm 2, \pm 3 \ldots$$

and $\qquad\qquad\qquad\qquad\qquad\qquad\qquad\qquad\qquad$ (13.35)

$$K_0 = \log|\det T(\varsigma)| = \text{Tr} \ \ell n \ T(\varsigma)$$

would be conserved.

Fundamental Poisson Bracket Relations:

As we have seen in Chapter 11, to prove integrability, it is essential to calculate the Poisson bracket of the Lax operator with itself. Furthermore, in the present case, the transition matrix, which contains all the scattering data, satisfies a Lax equation. The calculation of the Poisson bracket of $T(\zeta)$ with itself is, therefore, doubly important in the present case since it also would give the Poisson brackets between the scattering data which in turn would allow us to identify the action-angle variables. Consequently, this is what we discuss next.

Let us recall from Eq. (13.18) that in the example under study

$$A(x,\zeta) = i\sqrt{|\kappa|}\ (\psi^*\sigma_+ + \psi\sigma_-) - i\zeta\sigma_3 \qquad (13.36)$$

Consequently, using the relations in Eq. (12.45) and the tensor product notation introduced in Eqs. (11.41) and (11.42), we obtain

$$\{A(x,\zeta) \underset{,}{\otimes} A(y,\zeta')\} = i\kappa(\sigma_+ \otimes \sigma_- - \sigma_- \otimes \sigma_+)\delta(x-y) \qquad (13.37)$$

Let us next define the operator

$$P = \frac{1}{2}\left(I + \sum_{i=1}^{3} \sigma_i \otimes \sigma_i\right) \qquad (13.38)$$

This is quite analogous to the tensor Casimir operator for the group SU(2) except for the presence of the identity term. Using the commutation relations of the Pauli matrices, it can be shown easily that

$$[P \ , \ \sigma_j \otimes I + I \otimes \sigma_j] = 0$$

(13.39)

$$[P \ , \ \sigma_j \otimes \sigma_k + \sigma_k \otimes \sigma_j] = 0 \qquad \text{for } j,k = 1,2,3$$

This is interesting for we see that given any two arbitrary 2×2 matrices C and D, which can be expressed linearly in terms of the identity and the Pauli matrices, Eqs. (13.39) imply that

$$[P \ , \ C \otimes D + D \otimes C] = 0$$

$$\text{or,} \quad P(C \otimes D + D \otimes C) = (C \otimes D + D \otimes C)P \qquad (13.40)$$

Since P is nontrivial, it follows that for Eq. (13.40) to hold, we must have

$$P(C \otimes D) = (D \otimes C)P$$

(13.41)

In other words, P simply permutes the order of the tensor products and hence is also known as the permutation matrix. Its normalization is chosen so that

$$P^2 = I$$

(13.42)

From the definition of P in Eq. (13.38) and the relations in Eq. (13.40) and (13.41), it immediately follows that

$$[P , \psi^*(x)(\sigma_+ \otimes I + I \otimes \sigma_+)] = 0$$

$$[P , \psi(x)(\sigma_- \otimes I + I \otimes \sigma_-)] = 0 \tag{13.43}$$

Furthermore, using the commutation relations of the Pauli matrices, we can easily calculate to show that

$$[P , \sigma_3 \otimes I] = -[P , I \otimes \sigma_3]$$

$$= -2(\sigma_+ \otimes \sigma_- - \sigma_- \otimes \sigma_+) \tag{13.44}$$

We see from these relations, Eqs. (13.43)-(13.44), that

$$[P , A(x,\varsigma) \otimes I + I \otimes A(x,\varsigma')]$$

$$= 2i(\varsigma-\varsigma')(\sigma_+ \otimes \sigma_- - \sigma_- \otimes \sigma_+) \tag{13.45}$$

Comparing with Eq. (13.37), we recognize now that we can write

$$\{A(x,\varsigma) \underset{,}{\otimes} A(y,\varsigma')\}$$

$$= \delta(x-y)[r(\varsigma-\varsigma'),A(x,\varsigma) \otimes I + I \otimes A(x,\varsigma')] \tag{13.46}$$

where

$$r(\varsigma-\varsigma') = \frac{\kappa}{2(\varsigma-\varsigma')} \, P \qquad\qquad (13.47)$$

We see that Eq. (13.46) is the generalization of the relation in Eq. (11.54) to the present case and we note here that the r-matrix for our example depends on the spectral parameter. This is indeed a general feature of the first order formulation. We also recall from our discussion in Chapter 11, that a relation of the form in Eq. (13.46) plays a fundamental role in proving the integrability of the system and is, consequently, called the fundamental Poisson bracket relation.

Let us note from Eq. (13.46) that

$$r(\varsigma-\varsigma') = -r(\varsigma'-\varsigma) \qquad\qquad (13.48)$$

Using the above relation as well as Eq. (13.42), we can verify the antisymmetry of the fundamental Poisson bracket, which takes the form

$$\{A(x,\varsigma) \otimes A(y,\varsigma')\} = -P\{A(y,\varsigma') \otimes A(x,\varsigma)\}P \quad (13.49)$$

The fundamental Poisson bracket must also satisfy the Jacobi identity which for the tensor products takes the form

$$\{A(x,\varsigma) \otimes \{A(y,\varsigma') \otimes A(z,\varsigma'')\}\}$$

$$+ P_{23}P_{13}\{A(z,\varsigma'') \otimes \{A(x,\varsigma) \otimes A(y,\varsigma')\}\}P_{13}P_{23}$$

$$+ P_{12}P_{13}\{A(y,\varsigma') \otimes \{A(z,\varsigma'') \otimes A(x,\varsigma)\}\}P_{13}P_{12} = 0 \quad (13.50)$$

Let us note that each term in the Jacobi identity necessarily involves a triple tensor product space and P_{ij} simply permutes the ith and the jth embedded components in this tensor product space. We can substitute Eq. (13.46) into Eq. (13.50) and with an obvious generalization of the notation, we obtain the consistency condition to be

$$[r_{12}(\varsigma-\varsigma'),r_{13}(\varsigma)] + [r_{12}(\varsigma-\varsigma'),r_{23}(\varsigma')]$$

$$+ [r_{13}(\varsigma),r_{23}(\varsigma')] = 0 \qquad (13.51)$$

This relation is also known as the classical Yang-Baxter equation.

Action-Angle Variables:

Let us next note from the definition of $T(x,y;\varsigma)$ in Eq. (13.12) that we can use Eq. (13.46) to show that

$$\underset{\epsilon\to 0}{\text{Lim}} \{T(w+\epsilon,w-\epsilon;\varsigma) \otimes T(w+\epsilon,w-\epsilon;\varsigma')\}$$

$$= \underset{\epsilon\to 0}{\text{Lim}} [r(\varsigma-\varsigma') , T(w+\epsilon,w-\epsilon;\varsigma) \otimes T(w+\epsilon,w-\epsilon;\varsigma')] \quad (13.52)$$

Consequently,

$$\{T(x,y;\varsigma) \otimes T(x,y;\varsigma')\}$$

$$= \underset{\epsilon\to 0}{\text{Lim}} \underset{y<w<x}{\Pi} (T(x,w+\epsilon;\varsigma) \otimes T(x,w+\epsilon;\varsigma'))$$

$$\{T(w+\epsilon,w-\epsilon;\varsigma) \otimes T(w+\epsilon,w-\epsilon;\varsigma')\}$$

$$(T(w-\epsilon,y;\varsigma) \otimes T(w-\epsilon,y;\varsigma'))$$

and upon using Eq. (13.52), it follows that

$$\{T(x,y;\varsigma) \otimes T(x,y;\varsigma')\}$$

$$= [r(\varsigma-\varsigma') , T(x,y;\varsigma) \otimes T(x,y;\varsigma')] \quad (13.53)$$

Let us note here that although the matrix r defined in Eq. (13.47) is singular as $\varsigma \to \varsigma'$, the expressions on the right hand side of Eqs. (13.46), (13.52) and (13.53) are well behaved since the commutators vanish in that limit as can be seen from the properties of the permutation matrix in Eq. (13.41). Furthermore, for definiteness, let us choose the singularity behavior of the r-matrix to be principal value. This will be quite useful as we will see later.

Before calculating the Poisson bracket between two transition matrices, let us evaluate some expressions involving the r-matrix which would prove useful. Let us note from the permutation properties of the r-matrix that

$$\left(e^{i\varsigma\sigma_3 x} \otimes e^{i\varsigma'\sigma_3 x}\right) r(\varsigma-\varsigma') \left(e^{-i\varsigma\sigma_3 x} \otimes e^{-i\varsigma'\sigma_3 x}\right)$$

$$= r(\varsigma-\varsigma')\left(e^{-i(\varsigma-\varsigma')\sigma_3 x} \otimes e^{i(\varsigma-\varsigma')\sigma_3 x}\right)$$

Using the explicit form of $r(\varsigma-\varsigma')$ from Eqs. (13.38) and (13.47), the above expression simplifies to

$$\frac{\kappa}{2}\left[\frac{1}{2(\varsigma-\varsigma')}(I+\sigma_3 \otimes \sigma_3) + \frac{e^{-2i(\varsigma-\varsigma')x}}{(\varsigma-\varsigma')}\sigma_- \otimes \sigma_+\right.$$

$$\left. + \frac{e^{2i(\varsigma-\varsigma')x}}{(\varsigma-\varsigma')}\sigma_+ \otimes \sigma_-\right] \quad (13.54)$$

If we now use the formula

$$\lim_{x \to \pm\infty} \frac{e^{i\lambda x}}{\lambda} = \pm i\pi\delta(\lambda) \qquad (13.55)$$

we obtain

$$r_+(\varsigma - \varsigma') = \lim_{x \to \infty} \left(e^{i\varsigma\sigma_3 x} \otimes e^{i\varsigma'\sigma_3 x}\right)$$

$$r(\varsigma - \varsigma')\left(e^{-i\varsigma\sigma_3} \otimes e^{-i\varsigma'\sigma_3 x}\right)$$

$$= \frac{\kappa}{4(\varsigma - \varsigma')} \left(I + \sigma_3 \otimes \sigma_3\right)$$

$$+ \frac{i\pi\kappa}{2} \delta(\varsigma - \varsigma')(\sigma_+ \otimes \sigma_- - \sigma_- \otimes \sigma_+) \qquad (13.56)$$

and

$$r_-(\varsigma - \varsigma') = \lim_{x \to -\infty} \left(e^{i\varsigma\sigma_3 x} \otimes e^{i\varsigma'\sigma_3 x}\right)$$

$$r(\varsigma - \varsigma')\left(e^{-i\varsigma\sigma_3 x} \otimes e^{-i\varsigma'\sigma_3 x}\right)$$

$$= \frac{\kappa}{4(\varsigma - \varsigma')} \left(I + \sigma_3 \otimes \sigma_3\right)$$

$$- \frac{i\pi\kappa}{2} \delta(\varsigma - \varsigma')(\sigma_+ \otimes \sigma_- - \sigma_- \otimes \sigma_+) \qquad (13.57)$$

We are now in a position to calculate the Poisson bracket between the transition matrices. In fact, from the definition of $T(\varsigma)$ in Eq. (13.25) and using Eq. (13.53) we obtain

$$\{T(\varsigma) \underset{,}{\otimes} T(\varsigma')\} = \underset{\substack{x \to \infty \\ y \to -\infty}}{\text{Lim}} \left(e^{i\varsigma\sigma_3 x} \otimes e^{i\varsigma'\sigma_3 x}\right)$$

$$[r(\varsigma-\varsigma') , T(x,y;\varsigma) \otimes T(x,y;\varsigma')]$$

$$\left(e^{-i\varsigma\sigma_3 y} \otimes e^{-i\varsigma'\sigma_3 y}\right)$$

or, $\{T(\varsigma) \underset{,}{\otimes} T(\varsigma'\} = r_+(\varsigma-\varsigma')(T(\varsigma) \otimes T(\varsigma'))$

$$- (T(\varsigma) \otimes T(\varsigma'))r_-(\varsigma-\varsigma') \qquad (13.58)$$

where r_+ and r_- are defined in Eqs. (13.56) and (13.57) respectively. Explicitly, expression (13.58) takes the form

$$\{T(\varsigma) \underset{,}{\otimes} T(\varsigma')\} = \frac{\kappa}{4(\varsigma-\varsigma')} \left(\sigma_3 T(\varsigma) \otimes \sigma_3 T(\varsigma')\right.$$

$$\left. - T(\varsigma)\sigma_3 \otimes T(\varsigma')\sigma_3\right) + \frac{i\pi\kappa}{2} \delta(\varsigma-\varsigma')$$

$$\left(\sigma_+ T(\varsigma) \otimes \sigma_- T(\varsigma) + T(\varsigma)\sigma_+ \otimes T(\varsigma')\sigma_-\right.$$

$$\left. - \sigma_- T(\varsigma) \otimes \sigma_+ T(\varsigma') - T(\varsigma)\sigma_- \otimes T(\varsigma')\sigma_+\right) \qquad (13.59)$$

Equation (13.59) contains all the Poisson brackets between the scattering data. In fact, by taking the relevant matrix elements, we obtain

$$\{a(\varsigma) , a(\varsigma')\} = 0 = \{a(\varsigma) , a^*(\varsigma')\}$$

$$\{b(\varsigma) , b(\varsigma')\} = 0$$

$$\{b(\varsigma) , b^*(\varsigma')\} = -i\pi\kappa \, a(\varsigma)a^*(\varsigma)\delta(\varsigma-\varsigma') \qquad (13.60)$$

$$\{a(\varsigma) , b(\varsigma')\} = \frac{\kappa}{2(\varsigma-\varsigma'+io)} \, a(\varsigma)b(\varsigma')$$

$$\{a(\varsigma) , b^*(\varsigma')\} = - \frac{\kappa}{2(\varsigma-\varsigma'+io)} \, a(\varsigma)b^*(\varsigma')$$

It can be shown using Eq. (13.60) that if we define

$$Q(\varsigma) = \arg b(\varsigma)$$

$$\qquad (13.61)$$

$$P(\varsigma) = \frac{2}{\pi\kappa} \log|a(\varsigma)|$$

then

$$\{Q(\varsigma) , Q(\varsigma')\} = 0 = \{P(\varsigma) , P(\varsigma')\}$$

$$\qquad (13.62)$$

$$\{Q(\varsigma) , P(\varsigma')\} = \delta(\varsigma-\varsigma')$$

We can analytically continue the Poisson bracket relations in Eq. (13.60) to the upper half of the complex ς-plane and deduce for the discrete spectrum that the variables

$$q_n = 2 \log |b_n|$$

$$p_n = \frac{2}{\kappa} \, \text{Re} \, \varsigma_n$$

$$\qquad (13.63)$$

$$\tilde{q}_n = -2 \arg b_n$$

$$\tilde{p}_n = \frac{2}{\kappa} \, \text{Im} \, \varsigma_n$$

satisfy (see for example the book by Faddeev and Takhtajan)

$$\{q_n \ , \ p_m\} = \delta_{nm} = \{\tilde{q}_n \ , \ \tilde{p}_m\}$$

$$(13.64)$$

$$\{q_n \ , \ q_m\} = \{p_n \ , \ p_m\} = \{\tilde{q}_n \ , \ \tilde{q}_m\} = \{\tilde{p}_n \ , \ \tilde{p}_m\} = 0$$

Together the variables $Q(\varsigma)$, $P(\varsigma)$, q_n, p_n, $\tilde{q}_n$, $\tilde{p}_n$ defined in Eqs. (13.61) and (13.63) constitute the action angle variables of the system.

Let us recall from Eq. (5.54) that the analytic expression

$$\log a(\varsigma) = \frac{1}{\pi i} \int_{-\infty}^{\infty} d\varsigma' \ \frac{\log|a(\varsigma')|}{\varsigma' - \varsigma}$$

$$+ \sum_{m=1}^{N} \ \log\left(\frac{\varsigma - \varsigma_m}{\varsigma - \varsigma_m^*}\right) \qquad (13.65)$$

has the asymptotic expression given by

$$\log a(\varsigma) \xrightarrow[\varsigma \to \infty]{} \sum_{n=1}^{\infty} \frac{C_n}{\varsigma^n} \qquad (13.66)$$

where

$$c_n = -\frac{1}{\pi i} \int_{-\infty}^{\infty} d\varsigma \; \varsigma^{n-1} \log |a(\varsigma)|$$

$$+ \frac{1}{n} \sum_{m=1}^{N} \left((\varsigma_m^*)^n - \varsigma_m^n \right)$$

$$= -\frac{\kappa}{2i} \int_{-\infty}^{\infty} d\varsigma \; \varsigma^{n-1} P(\varsigma)$$

$$+ \frac{1}{n} (\frac{\kappa}{2})^n \sum_{m=1}^{N} \left((p_m - i\tilde{p}_m)^n - (p_m + i\tilde{p}_m)^n \right) \quad (13.67)$$

Let us note here that the asymptotic expansion of $\log a(\varsigma)$ involves terms depending only on the action variables and consequently, each term in the expansion is in involution with the others.

From our discussion in Chapter 5 (see Eqs. (5.56)-(5.59)), we recall that we can obtain the asymptotic expansion of $\log a(\varsigma)$ alternately in the following way. In terms of the Jost function in Eq. (12.63), let us define

$$\xi(x,\varsigma) = \frac{1}{f_2(x,\varsigma)} \frac{\partial f_2(x,\varsigma)}{\partial x} - i\varsigma \quad (13.68)$$

where $f_2(x,\varsigma)$ is the second component of the Jost function $f(x,\varsigma)$. As in Chapter 5, we can show using the properties of the Jost functions that (see Eq. (5.59))

$$\log a(\varsigma) = -\int_{-\infty}^{\infty} dx \; \xi(x,\varsigma) \quad (13.69)$$

Furthermore, we see from Eq. (12.48) that $f_2(x,\varsigma)$ satisfies the second order equation given by

$$\frac{\partial^2 f_2(x,\varsigma)}{\partial x^2} + (\varsigma^2 - \kappa\psi^*\psi)f_2(x,\varsigma)$$

$$- i\sqrt{|\kappa|}\,\frac{\partial\psi}{\partial x}\,f_1(x,\varsigma) = 0 \qquad (13.70)$$

It follows, therefore, that $\xi(x,\varsigma)$ satisfies the relation

$$\psi\,\frac{\partial}{\partial x}\left(\frac{\xi}{\psi}\right) + 2i\varsigma\xi + \xi^2 - \kappa\psi^*\psi = 0 \qquad (13.71)$$

If we now assume the asymptotic expansion

$$\xi(x,\varsigma)\xrightarrow[\varsigma\to\infty]{}\sum_{n=1}^{\infty}\frac{\xi_n(x)}{(2i\varsigma)^n} \qquad (13.72)$$

then Eq. (13.71) implies a recursion relation between the coefficient functions of the form

$$\xi_{n+1} + \psi\,\frac{\partial}{\partial x}\left(\frac{\xi_n}{\psi}\right) + \sum_{m=1}^{n}\xi_{n-m}\xi_m = 0 \qquad n = 1,2,3\ldots \quad (13.73)$$

with

$$\xi_1 = \kappa\psi^*\psi$$

From Eq. (13.69) we see that log a(ς) has the asymptotic expansion, in terms of these coefficient function, given by

$$\log a(\varsigma) \longrightarrow - \sum_{n=1}^{\infty} \frac{1}{(2i\varsigma)^n} \int_{-\infty}^{\infty} dx \, \xi_n(x) \qquad (13.74)$$

Comparing Eqs. (13.66) and (13.74), we obtain

$$\int_{-\infty}^{\infty} dx \, \xi_n(x) = -(2i)^n \, c_n \qquad (13.75)$$

We recall from Eq. (13.35) that log a(ς) is conserved and, consequently, each term in its asymptotic expansion is also conserved. We see from Eqs. (13.66) and (13.74) that there is an infinite number of conserved quantities and they are in involution since the c_n's are in involution as we have noted earlier. Consequently, the system under study, namely, the nonlinear Schrödinger equation with $\kappa < 0$ is integrable. We also note here that Eq. (13.75) provides a map of the conserved quantities expressed in terms of the dynamical variables ψ, $\psi*$ to those expressed in terms of the action variables.

Let us next calculate the first few conserved quantities from Eq. (13.73).

$$H_1 = \int_{-\infty}^{\infty} dx \, \xi_1(x) = \kappa \int_{-\infty}^{\infty} dx \, \psi*\psi \qquad (13.76)$$

$$H_2 = \int_{-\infty}^{\infty} dx \, \xi_2(x) = \frac{\kappa}{2} \int_{-\infty}^{\infty} dx \left(\psi* \frac{\partial \psi}{\partial x} - \frac{\partial \psi*}{\partial x} \psi\right) \qquad (13.77)$$

$$H_3 = -\kappa \int_{-\infty}^{\infty} dx \left(\frac{\partial \psi^*}{\partial x} \frac{\partial \psi}{\partial x} + \kappa \; (\psi^*\psi)^2 \right) \qquad (13.78)$$

We recognize H_1 to be the number number of particles up to a multiplicative constant κ. Similarly H_2 represents the momentum and H_3 the energy of the nonlinear Schrödinger equation (up to a constant) given in Eq. (12.44). In fact, let us note that

$$H = -\frac{1}{\kappa} H_3 = \frac{(2i)^3}{\kappa} C_3 \qquad (13.79)$$

Using Eq. (13.67), we see that in terms of the action variables, the Hamiltonian becomes

$$H = 4 \int_{-\infty}^{\infty} d\varsigma \; \varsigma^2 P(\varsigma) + 2\kappa^2 \sum_{m=1}^{N} \left(\frac{1}{3} \tilde{p}_m^3 - p_m^2 \tilde{p}_m \right) \qquad (13.80)$$

We can now evaluate the time evolution of various quantities.

$$\dot{Q}(\varsigma) = \{Q(\varsigma) \; , \; H\} = 4\varsigma^2$$

$$\dot{q}_n = \{q_n \; , \; H\} = -4\kappa^2 p_n \tilde{p}_n = -16 \; \mathrm{Re} \; \varsigma_n \; \mathrm{Im} \; \varsigma_n \qquad (13.81)$$

$$\dot{\tilde{q}}_n = \{\tilde{q}_n \; , \; H\} = 2\kappa^2 \left(\tilde{p}_n^2 - p_n^2 \right) = -8 \left(\mathrm{Re} \; \varsigma_n^2 - \mathrm{Im} \; \varsigma_n^2 \right)$$

These relations can be equivalently written as

$$b(\varsigma,t) = b(\varsigma,0) \, e^{4i\varsigma^2 t}$$

$$(13.82)$$

$$b_n(t) = b_n(0) \, e^{4i\varsigma_n^2 t}$$

We recognize these as the same relations we had obtained in Eq. (12.84).

References:

Faddeev, L. D., Sov. Sci. Reviews C1, Ed. S. P. Novikov, 1981.

Faddeev, L. D. in Recent Advances in Field Theory and Statistical Mechanics, Les Houches Summer School Proc., Vol. 39, North Holland, 1984.

Faddeev, L. D. in Lecture Notes Phys., Vol. 242, Ed. B. S. Shastry et al, Springer-Verlag, 1985.

Faddeev, L. D. and L. A. Takhtajan, Hamiltonian Methods in the Theory of Solitons, Springer-Verlag, 1987.

Kulish, P. P. and E. K. Sklyanin in Lect. Notes Phys. Vol. 151, Ed. J. Hietarinta et al, Springer-Verlag 1982.

Thacker, H. B., Rev. Mod. Phys. 53, 253 (1981).

Thacker, H. B. in Lect. Notes Phys., Vol. 151, Ed. J. Hietarinta et al, Springer-Verlag, 1982.

Sklyanin, E. K., J. Sov. Math. 19, 1546 (1982).

Takhtadzhyan, L. A. and L. D. Faddeev, Theo. Math. Phys. 21, 1046 (1975).

Zakharov, V. E. and A. B. Shabat, JETP 34, 62 (1972).

Zakharov, V. E., L. A. Takhtadzhyan and L. D. Faddeev, Sov. Phys. Dokl. 19, 824 (1975).

CHAPTER 14

QUANTUM INTEGRABILITY

Our discussion of integrable systems has so far been at the classical level. In a quantum theory, of course, the dynamical variables must be quantized as operators and various quantum commutation relations must take the place of Poisson brackets. Furthermore, since the quantized variables do not commute, we must address the question of operator ordering and wherever necessary, the quantum expressions must be regularized.

The traditional approach to quantum integrable systems is known as the Bethe ansatz method. Furthermore, as we have emphasized repeatedly, the inverse scattering method generalizes readily to quantum systems and can also be used to study quantum integrable systems. In this chapter, we will discuss the Bethe ansatz method as well as the quantum inverse scattering method and point out the relation between the two approaches. Finally, we will conclude with a brief discussion of the Yang-Baxter equation and quantum groups.

The Bethe Ansatz:

By integrability of a quantum system we understand that we can determine the spectrum of the Hamiltonian as well as its scattering matrix. Traditionally, the diagonalization of the Hamiltonian is achieved, in most one dimensional integrable systems, through the Bethe ansatz - named after Bethe who had first made this ansatz for the wave function in his study of a lattice model of the magnets.

Let us study the ansatz in some detail in the particular example of the nonlinear Schrödinger system. For simplicity, we will assume here a repulsive potential, namely, $\kappa > 0$. (This eliminates the existence of bound states.) In the quantum theory the dynamical variables $\psi(x,t)$ and $\psi^*(x,t)$ become the operators $\hat{\psi}(x,t)$ and $\hat{\psi}^\dagger(x,t)$ respectively where $\dagger$ stands for Hermitian conjugation. From the correspondence between the Poisson brackets and the quantum commutators, we see that the quantization conditions for the operators $\hat{\psi}$ and $\hat{\psi}^\dagger$ can be obtained from Eq. (12.45) to be

$$[\hat{\psi}(x,t) \, , \, \hat{\psi}^\dagger(y,t)] = \hbar\delta(x-y)$$

$$(14.1)$$

$$[\hat{\psi}(x,t) \, , \, \hat{\psi}(y,t)] = 0 = [\hat{\psi}^\dagger(x,t) \, , \, \hat{\psi}^\dagger(y,t)]$$

where $\hbar$ is the Planck's constant.

Let us note here that the quantum nonlinear Schrödinger equation is a relatively simple system to study since it can be regularized completely if we normal order quantum expressions using Eq. (14.1) so that all the $\hat{\psi}$ operators stand to the right of $\hat{\psi}^\dagger$. For example, note from Eq. (12.44) that the normal ordered quantum Hamiltonian for the system takes the form

$$\hat{H} = \int_{-\infty}^{\infty} dx \left[\hat{\psi}^\dagger_x\hat{\psi}_x + \kappa\hat{\psi}^\dagger\hat{\psi}^\dagger\hat{\psi}\hat{\psi}\right]$$

$$(14.2)$$

It follows, then, that the Heisenberg equations of motion are given by

$$\hat{\psi}_t = \frac{1}{i\hbar} \, [\hat{\psi}, H]$$

which upon using Eq. (14.1) becomes

$$i\hat{\psi}_t = -\hat{\psi}_{xx} + 2\kappa\hat{\psi}^\dagger\hat{\psi}\hat{\psi} \qquad\qquad (14.3)$$

We recognize this (see Eq. (12.46)) to be the quantum nonlinear Schrödinger equation.

Since $\hat{\psi}$ and $\hat{\psi}^\dagger$ satisfy the canonical commutation relations even in the presence of interactions, we can build up the Hilbert space of the system in the following way. Let us define the vacuum state $|0\rangle$ which satisfies

$$\hat{\psi}(x,t)|0\rangle = 0 \qquad\qquad (14.4)$$

Then we can think of $\hat{\psi}^\dagger$ as the creation operator and write the N-particle state with momenta $k_1, k_2 \ldots k_N$ as (for the rest of this section, we are going to assume $\hbar = 1$ for simplicity)

$$|k_1 \ldots k_N\rangle$$

$$= \frac{1}{\sqrt{N!}} \int_{-\infty}^{\infty} dx_1 \ldots dx_N \phi(x_1 \ldots x_N; k_1 \ldots k_N)\hat{\psi}^\dagger(x_1) \ldots \hat{\psi}^\dagger(x_N)|0\rangle \quad (14.5)$$

where

$$\phi(x_1 \ldots x_N; k_1 \ldots k_N) = \langle 0 | \hat{\psi}(x_1) \ldots \hat{\psi}(x_N) | k_1 \ldots k_N \rangle \qquad (14.6)$$

is the N-particle wave function of the system. For a noninteracting system, the wave function would simply correspond to plane waves. However, since the nonlinear Schrödinger equation describes an interacting system, the wave function would have more structure and Bethe's ansatz determines its form.

Let us note from Eqs. (14.1), (14.2) and (14.4) that we can write

$$\hat{H} | k_1 \ldots k_N \rangle$$

$$= \frac{1}{\sqrt{N!}} \int_{-\infty}^{\infty} dx_1 \ldots dx_N \Big[\Big(- \sum_{i=1}^{N} \frac{\partial^2}{\partial x_i^2} + 2\kappa \sum_{\substack{i,j=1 \\ i<j}}^{N} \delta(x_i - x_j) \Big)$$

$$\phi(x_1 \ldots x_N; k_1 \ldots k_N) \Big] \hat{\psi}^\dagger(x_1) \ldots \hat{\psi}^\dagger(x_N) | 0 \rangle \quad (14.7)$$

It is obvious, therefore, that $| k_1 \ldots k_N \rangle$ is an eigenstate of the total Hamiltonian if the wave function ϕ is an eigenfunction of the many-body Schrödinger equation with δ-function interactions. This is how the study of the quantum nonlinear Schrödinger equation becomes equivalent to the study of a Bose gas with point interactions. To understand the structure of the wave function in this case, let us analyze the two-particle wave function in some detail. The eigenvalue equation, we are interested in, is given by

$$\left(- \frac{\partial^2}{\partial x_1^2} - \frac{\partial^2}{\partial x_2^2} + 2\kappa\delta(x_1 - x_2)\right)\phi(x_1, x_2; k_1, k_2)$$

$$= \lambda\phi(x_1, x_2; k_1, k_2) \qquad (14.8)$$

We recall that a δ-function potential in the Schrödinger equation leads to a discontinuity in the first derivative of the wave function at the interaction points. Taking care of this appropriately, we can check easily that the solution of Eq. (14.8) takes the form

$$\phi(x_1, x_2; k_1, k_2) = e^{i(k_1 x_1 + k_2 x_2)}\left(1 - \frac{2i\kappa}{k_2 - k_1}\,\epsilon(x_2 - x_1)\right)$$

$$+ e^{i(k_2 x_1 + k_1 x_2)}\left(1 - \frac{2i\kappa}{k_1 - k_2}\,\epsilon(x_2 - x_1)\right) \quad (14.9)$$

where $\epsilon(x)$ is the alternating step function defined in Eq. (1.59). The eigenvalue λ in this case can be shown to equal

$$\lambda = k_1^2 + k_2^2 \qquad (14.10)$$

We see that in the region of space where $x_1 < x_2$ the wave function takes the simple form (recall that $\epsilon(x) = (\theta(x) - 1/2)$)

$$\phi(x_1, x_2; k_1, k_2) = \frac{k_2 - k_1 - i\kappa}{k_2 - k_1}\, e^{i(k_1 x_1 + k_2 x_2)} +$$

$$+ \frac{k_1 - k_2 - i\kappa}{k_1 - k_2} e^{i(k_2 x_1 + k_1 x_2)}$$

$$= \sum_P e^{i(k_{P_1} x_1 + k_{P_2} x_2)} \frac{k_{P_2} - k_{P_1} - i\kappa}{k_{P_2} - k_{P_1}}$$

where P_1 and P_2 represent the two permutations of the numbers 1 and 2 and the summation is over the two permutations.

Bethe ansatz consists of generalizing the structure of this wave function to a N-particle wave function. In the region $x_1 < x_2 \ldots < x_N$, the Bethe ansatz for the wave function takes the form

$$\phi(x_1, \ldots x_N; k_1 \ldots k_N)$$

$$= \sum_P e^{i(k_{P_1} x_1 + \ldots + k_{P_N} x_N)} \prod_{i > j} \frac{k_{P_i} - k_{P_j} - i\kappa}{k_{P_i} - k_{P_j}} \qquad (14.11)$$

where $P_1, \ldots . P_N$ denote the permutations of the numbers $1, 2, \ldots N$ and the summation is over all such permutations. This wave function can be readily extended to the entire coordinate space and can be shown to be the eigenfunction of the N-body Schrödinger Hamiltonian on the right hand side of Eq. (14.7) with the eigenvalue $\sum_{i=1}^{N} k_i^2$. Thus we see that with these wave functions, the N-particle state defined in Eq. (14.5) satisfies

$$\hat{H}|k_1,k_2\ldots k_N\rangle = \Big(\sum_{i=1}^{N} k_i^2\Big)|k_1,k_2\ldots k_N\rangle \qquad (14.12)$$

In other words, these many particle states are the exact eigenstates of the full Hamiltonian.

Quantum Inverse Scattering:

As we have emphasized repeatedly, the method of inverse scattering carries over readily to the quantum system. In particular, most of the ideas developed in the last chapter as well as the relations obtained there generalize to the quantum system once we take care of the operator nature of the dynamical variables. The operator character also introduces new structures into the theory and in this section we will discuss these ideas in some detail.

The auxiliary linear problem of Eq. (13.13) takes a normal ordered form in the quantum case, namely, (recall that Eq. (13.13) holds for $\kappa < 0$)

$$\frac{\partial\hat{\phi}(x)}{\partial x} = :\hat{A}(x,\varsigma)\hat{\phi}(x):$$

$$= -i\varsigma\sigma_3\hat{\phi}(x) + i\sqrt{|\kappa|}\hat{\psi}^\dagger(x)\sigma_+\hat{\phi}(x)$$

$$+ i\sqrt{|\kappa|}\sigma_-\hat{\phi}(x)\hat{\psi}(x) \qquad (14.13)$$

where the solutions are functionals of the dynamical variables and, consequently, are defined to be normal ordered, that is,

$$\hat{\phi}(x) = :\phi(x): \qquad\qquad (14.14)$$

Let us note here that even though the Lax operator is a quantized operator in the quantum theory, the relation of Eq. (13.45) still holds since P is a c-number matrix operator. In other words,

$$[r(\varsigma-\varsigma'),\hat{A}(x,\varsigma) \otimes I + I \otimes \hat{A}(x,\varsigma')]$$

$$= \frac{\kappa}{2(\varsigma-\varsigma')} [P,\hat{A}(x,\varsigma) \otimes I + I \otimes \hat{A}(x,\varsigma')]$$

$$= i\kappa(\sigma_+ \otimes \sigma_- - \sigma_- \otimes \sigma_+) \qquad (14.15)$$

It follows, from this as well as from various properties of the r-matrix (see Eq. (13.41) in particular), that

$$\left(I - i\hbar r(\varsigma-\varsigma')\right)\left(\hat{A}(x,\varsigma) \otimes I + I \otimes \hat{A}(x,\varsigma') + \hbar\kappa\sigma_- \otimes \sigma_+\right)$$

$$= \left(\hat{A}(x,\varsigma) \otimes I + I \otimes \hat{A}(x,\varsigma') + \hbar\kappa\sigma_+ \otimes \sigma_-\right)\left(I - i\hbar r(\varsigma-\varsigma')\right)$$

or, $R(\varsigma-\varsigma')\left(\hat{A}(x,\varsigma) \otimes I + I \otimes \hat{A}(x,\varsigma') + \hbar\kappa\sigma_- \otimes \sigma_+\right)$

$$= \left(\hat{A}(x,\varsigma) \otimes I + I \otimes \hat{A}(x,\varsigma') + \hbar\kappa\sigma_+ \otimes \sigma_-\right)R(\varsigma-\varsigma') \quad (14.16)$$

where we have defined

$$R(\varsigma-\varsigma') = I - i\hbar r(\varsigma-\varsigma') \qquad\qquad (14.17)$$

Several comments are in order here. We have introduced the Planck's constant to emphasize the quantum nature of the relations. Let us note that R is till a c-number matrix even though it depends on the quantum parameter. Furthermore, Eq. (14.16) is the quantum analogue of the classical relation in Eq. (13.46) and plays an important role in the study of quantum integrability. We also note here that Eq. (14.16) reduces to Eq. (13.46) in the classical limit according to the correspondence principle

$$\frac{1}{i\hbar} [\quad , \quad] \xrightarrow[\hbar \to 0]{} \{ \quad , \quad \} \tag{14.18}$$

Let us next study the properties of the quantum $T(x,y;\varsigma)$ matrix. By definition,

$$\hat{T}(x,y;\varsigma) = :T(x,y;\varsigma): = :e^{\int_{y}^{x} dz \, \hat{A}(z,\varsigma)}: \tag{14.19}$$

It is obvious that

$$\hat{T}(x,x;\varsigma) = I \tag{14.20}$$

However, because the dynamical variables no longer commute, the semigroup property in Eq. (13.15) now becomes

$$\hat{T}(x,y;\varsigma)\hat{T}(y,z;\varsigma) = \hat{T}(x,z;\varsigma) \quad \text{for} \quad x > y > z \tag{14.21}$$

The differential equations satisfied by $\hat{T}$ now take the form

$$\frac{\partial \hat{T}(x,y;\varsigma)}{\partial x} = :\hat{A}(x,\varsigma)\hat{T}(x,y;\varsigma):$$

$$= -i\varsigma\sigma_3\hat{T}(x,y;\varsigma) + i\sqrt{|\kappa|}\ \hat{\psi}^\dagger(x)\sigma_+\hat{T}(x,y;\varsigma)$$

$$+ i\sqrt{|\kappa|}\ \sigma_-\hat{T}(x,y;\varsigma)\hat{\psi}(x) \qquad (14.22)$$

and

$$\frac{\partial \hat{T}(x,y;\varsigma)}{\partial y} = -:\hat{T}(x,y;\varsigma)\hat{A}(y,\varsigma):$$

$$= i\varsigma\hat{T}(x,y;\varsigma)\sigma_3 - i\sqrt{|\kappa|}\ \hat{\psi}^\dagger(y)\hat{T}(x,y;\varsigma)\sigma_+$$

$$- i\sqrt{|\kappa|}\ \hat{T}(x,y;\varsigma)\hat{\psi}(y)\sigma_- \qquad (14.23)$$

Note that the differential equations (14.22) and (14.23) can
be written in the equivalent integral forms as

$$\hat{T}(x,y;\varsigma) = I + \int_y^x dz:\hat{A}(z,\varsigma)\hat{T}(z,y;\varsigma): \qquad (14.24)$$

$$\hat{T}(x,y;\varsigma) = I + \int_y^x dz:\hat{T}(x,z;\varsigma)\hat{A}(z,\varsigma): \qquad (14.25)$$

Using the form of $\hat{A}$ as well as the commutation relations of
Eq. (14.1) we see that we can write

$$[\hat{\psi}(x),\hat{T}(x,y;\varsigma)] = \frac{i\hbar\sqrt{|\kappa|}}{2} \sigma_+\hat{T}(x,y;\varsigma)$$

$$[\hat{\psi}^\dagger(x),\hat{T}(x,y;\varsigma)] = - \frac{i\hbar\sqrt{|\kappa|}}{2} \sigma_-\hat{T}(x,y;\varsigma)$$

$$(14.26)$$

$$[\hat{\psi}(y),\hat{T}(x,y;\varsigma)] = \frac{i\hbar\sqrt{|\kappa|}}{2} \hat{T}(x,y;\varsigma)\sigma_+$$

$$[\hat{\psi}^\dagger(y),\hat{T}(x,y;\varsigma)] = - \frac{i\hbar\sqrt{|\kappa|}}{2} \hat{T}(x,y;\varsigma)\sigma_-$$

where we have used

$$\int_y^x dz\ \delta(x-z) = \frac{1}{2} = \int_y^x dz\ \delta(y-z) \qquad (14.27)$$

Let us next define the following two products of the $\hat{T}$ operators. Let

$$\hat{T}_1(x,y;\varsigma) = \hat{T}(x,y;\varsigma) \otimes I$$

$$(14.28)$$

$$\hat{T}_2(x,y;\varsigma') = I \otimes \hat{T}(x,y;\varsigma')$$

Namely, $\hat{T}_1$ and $\hat{T}_2$ represent particular embeddings of the $\hat{T}$ operator into the tensor product space. It is clear that classically

$$\hat{T}_1(x,y;\varsigma)\hat{T}_2(x,y;\varsigma') = \left(\hat{T}(x,y;\varsigma) \otimes I\right)\left(I \otimes \hat{T}(x,y;\varsigma')\right)$$

$$= \hat{T}(x,y;\varsigma) \otimes \hat{T}(x,y;\varsigma')$$

and

$$\hat{T}_2(x,y;\varsigma')\hat{T}_1(x,y;\varsigma) = \left(I \otimes \hat{T}(x,y;\varsigma')\right)\left(\hat{T}(x,y;\varsigma) \otimes I\right)$$

$$= \hat{T}(x,y;\varsigma) \otimes \hat{T}(x,y;\varsigma')$$

so that, classically,

$$\hat{T}_1(x,y;\varsigma)\hat{T}_2(x,y;\varsigma') = \hat{T}_2(x,y;\varsigma')\hat{T}_1(x,y;\varsigma) \qquad (14.29)$$

In the quantum theory, however, operators do not commute in general and hence the equality in Eq. (14.29) does not hold. Finding a relation between the two products in Eq. (14.29), therefore, would determine the quantum noncommutativity and hence would lead to the commutators between the scattering data operators. This is what we do next.

Let us note from Eq. (14.22) that

$$\frac{\partial}{\partial x}\left(\hat{T}_1(x,y;\varsigma)\hat{T}_2(x,y;\varsigma')\right)$$

$$= :\hat{A}(x,\varsigma)\hat{T}_1(x,y;\varsigma):\hat{T}_2(x,y;\varsigma') + \hat{T}_1(x,y;\varsigma):\hat{A}(x,\varsigma')\hat{T}_2(x,y;\varsigma'):$$

$$= \left(-i\varsigma(\sigma_3 \otimes I) - i\varsigma'(I \otimes \sigma_3)\right)\hat{T}_1(x,y;\varsigma)\hat{T}_2(x,y;\varsigma')$$

$$+ i\sqrt{|\kappa|}\left(\sigma_+ \otimes I + I \otimes \sigma_+\right)\hat{\psi}^\dagger(x)\hat{T}_1(x,y;\varsigma)\hat{T}_2(x,y;\varsigma') -$$

$$- i\sqrt{|\kappa|} \; (I \otimes \sigma_+) \left[\hat{\psi}^{\dagger}(x), \hat{T}_1(x,y;\varsigma) \right] \hat{T}_2(x,y;\varsigma')$$

$$+ i\sqrt{|\kappa|} \; (\sigma_- \otimes I + I \otimes \sigma_-) \hat{T}_1(x,y;\varsigma) \hat{T}_2(x,y;\varsigma') \hat{\psi}(x)$$

$$+ i\sqrt{|\kappa|} \; (\sigma_- \otimes I) \hat{T}_1(x,y;\varsigma) \left[\hat{\psi}(x), \hat{T}_2(x,y;\varsigma') \right] \quad (14.30)$$

If we now introduce the operators

$$\hat{L}_1(x,\varsigma,\varsigma') = \hat{A}(x,\varsigma) \otimes I + I \otimes \hat{A}(x,\varsigma') + \hbar\kappa\sigma_- \otimes \sigma_+$$

$$(14.31)$$

$$\hat{L}_2(x,\varsigma,\varsigma') = \hat{A}(x,\varsigma) \otimes I + I \otimes \hat{A}(x,\varsigma') + \hbar\kappa\sigma_+ \otimes \sigma_-$$

and use the commutation relations in Eq. (14.26), then Eq. (14.30) simplifies to

$$\frac{\partial}{\partial x} \left(\hat{T}_1(x,y;\varsigma) \hat{T}_2(x,y;\varsigma') \right)$$

$$= \; :\hat{L}_1(x,\varsigma,\varsigma') \hat{T}_1(x,y;\varsigma) \hat{T}_2(x,y;\varsigma'): \quad (14.32)$$

In deriving this result, we have used the fact that in our analysis κ is negative. We can, similarly, show that

$$\frac{\partial}{\partial x} \left(\hat{T}_2(x,y;\varsigma') \hat{T}_1(x,y;\varsigma) \right)$$

$$= \; :\hat{L}_2(x,\varsigma,\varsigma') \hat{T}_2(x,y;\varsigma') \hat{T}_1(x,y;\varsigma): \quad (14.33)$$

We recall from the defining relation in Eq. (14.17) that R is a constant matrix. Furthermore, from Eq. (14.16) we see that

$$R(\varsigma-\varsigma')\hat{L}_1(x,\varsigma,\varsigma') = \hat{L}_2(x,\varsigma,\varsigma')R(\varsigma-\varsigma') \qquad (14.34)$$

Using these we can derive from Eqs. (14.32) and (14.33) that

$$R(\varsigma-\varsigma')\hat{T}_1(x,y;\varsigma)\hat{T}_2(x,y;\varsigma')$$

$$= \hat{T}_2(x,y;\varsigma')\hat{T}_1(x,y;\varsigma)R(\varsigma-\varsigma') \qquad (14.35)$$

This relation gives the quantum noncommutativity of the $\hat{T}$ operators. It is a relation of fundamental importance in the study of quantum integrability and is the quantum analogue of Eq. (13.53). Indeed, in the classical limit, Eq. (14.35) reduces to Eq. (13.53) as can be checked directly.

To calculate the quantum commutation relation between the transition matrices, let us note from Eqs. (14.32) and (14.33) that the asymptotic behaviors of the product functions are given by

$$\lim_{y\to-\infty} \hat{T}_1(x,y;\varsigma)\hat{T}_2(x,y;\varsigma') \longrightarrow e^{\hat{L}_1^{(0)}x}$$

$$\lim_{x\to\infty} \hat{T}_1(x,y;\varsigma)\hat{T}_2(x,y;\varsigma') \longrightarrow e^{\hat{L}_1^{(0)}y}$$

$$(14.36)$$

$$\lim_{y\to-\infty} \hat{T}_2(x,y;\varsigma')\hat{T}_1(x,y;\varsigma) \longrightarrow e^{\hat{L}_2^{(0)}x}$$

$$\lim_{x\to\infty} \hat{T}_2(x,y;\varsigma')\hat{T}_1(x,y;\varsigma) \longrightarrow e^{\hat{L}_2^{(0)}y}$$

where

$$\hat{L}_1^{(0)} = \lim_{|x| \to \infty} \hat{L}_1(x, \varsigma, \varsigma')$$

$$\hat{L}_2^{(0)} = \lim_{|x| \to \infty} \hat{L}_2(x, \varsigma, \varsigma')$$

(14.37)

These asymptotic behaviors are clearly different from the classical case and consequently, the limiting process to obtain the products of the transition matrices must be taken carefully. A careful analysis leads to the result

$$R(\varsigma - \varsigma')\left(I + \frac{i\hbar\kappa}{2(\varsigma - \varsigma' - io)} \sigma_- \otimes \sigma_+\right)$$

$$\hat{T}_1(\varsigma)\hat{T}_2(\varsigma')\left(I - \frac{i\hbar\kappa}{2(\varsigma - \varsigma' + io)} \sigma_- \otimes \sigma_+\right)$$

$$= \left(I - \frac{i\hbar\kappa}{2(\varsigma - \varsigma' + io)} \sigma_+ \otimes \sigma_-\right)$$

$$\hat{T}_2(\varsigma')\hat{T}_1(\varsigma)\left(I + \frac{i\hbar\kappa}{2(\varsigma - \varsigma' - io)} \sigma_+ \otimes \sigma_-\right)R(\varsigma - \varsigma') \quad (14.38)$$

This is the quantum analogue of Eq. (13.59). If we now define the quantities

$$R_\pm(\varsigma - \varsigma') = \left(I + \frac{i\hbar\kappa}{2(\varsigma - \varsigma' \pm io)} \sigma_+ \otimes \sigma_-\right)$$

$$R(\varsigma - \varsigma')\left(I + \frac{i\hbar\kappa}{2(\varsigma - \varsigma' \mp io)} \sigma_- \otimes \sigma_+\right)$$

(14.39)

then relation (14.38) can be written in the simple form

$$R_+(\varsigma-\varsigma')\hat{T}_1(\varsigma)\hat{T}_2(\varsigma') = \hat{T}_2(\varsigma')\hat{T}_1(\varsigma)R_-(\varsigma-\varsigma') \quad (14.40)$$

If we further recall the form of the transition matrix, namely,

$$\hat{T}(\varsigma) = \begin{pmatrix} \hat{a}(\varsigma) & \hat{b}(\varsigma) \\ -\hat{b}^\dagger(\varsigma) & \hat{a}^\dagger(\varsigma) \end{pmatrix} \quad (14.41)$$

then the quantum commutation relations can be read out from Eq. (14.40) by taking appropriate matrix elements. They take the form

$$[\hat{a}(\varsigma),\hat{a}(\varsigma')] = 0 = [\hat{a}(\varsigma),\hat{a}^\dagger(\varsigma')]$$

$$[\hat{b}(\varsigma),\hat{b}(\varsigma')] = 0$$

$$[\hat{a}(\varsigma),\hat{b}(\varsigma')] = \frac{i\hbar\kappa}{2(\varsigma-\varsigma'+io)} \hat{b}(\varsigma')\hat{a}(\varsigma) \quad (14.42)$$

$$[\hat{a}(\varsigma),\hat{b}^\dagger(\varsigma')] = -\frac{i\hbar\kappa}{2(\varsigma-\varsigma'+io)} \hat{a}(\varsigma)\hat{b}^\dagger(\varsigma')$$

$$[\hat{b}(\varsigma),\hat{b}^\dagger(\varsigma')] = \pi\hbar\kappa\delta(\varsigma-\varsigma')\hat{a}(\varsigma)\hat{a}^\dagger(\varsigma)$$

$$- \frac{\hbar^2\kappa^2}{4} \left(\frac{1}{(\varsigma-\varsigma')(\varsigma-\varsigma'-io)} - \frac{i\pi\delta(\varsigma-\varsigma')}{\varsigma-\varsigma'+io}\right)\hat{b}(\varsigma)\hat{b}^\dagger(\varsigma')$$

Note that in the limit of $\hbar \to 0$, the relations in Eq. (14.42) reduce to the classical relations of Eq. (13.60). Furthermore, it is obvious that

$$[\log \hat{a}(\varsigma) \ , \ \log \hat{a}(\varsigma')] = 0 \qquad (14.43)$$

so that as in the classical case, $\log \hat{a}(\varsigma)$ generates commuting quantities. The Hamiltonian can be shown to be contained in $\log \hat{a}(\varsigma)$ so that these commuting quantities are conserved also.

Let us next define the operators

$$\hat{\phi}(\varsigma) = \left(\pi \hbar |\kappa| \hat{a}^\dagger(\varsigma) \hat{a}(\varsigma)\right)^{-1/2} \hat{b}^\dagger(\varsigma)$$

$$\hat{\phi}^\dagger(\varsigma) = \hat{b}(\varsigma) \left(\pi \hbar |\kappa| \hat{a}^\dagger(\varsigma) \hat{a}(\varsigma)\right)^{-1/2} \qquad (14.44)$$

Then using the commutation relations in Eq. (14.42) we can show that

$$\left[\hat{\phi}(\varsigma), \hat{\phi}(\varsigma')\right] = 0 = \left[\hat{\phi}^\dagger(\varsigma), \hat{\phi}^\dagger(\varsigma')\right]$$

$$\left[\hat{\phi}(\varsigma), \hat{\phi}^\dagger(\varsigma')\right] = \delta(\varsigma - \varsigma') \qquad (14.45)$$

as well as

$$\left[\log \hat{a}(\varsigma), \hat{\phi}^\dagger(\varsigma')\right] = \log \left(1 + \frac{i \hbar \kappa}{2(\varsigma - \varsigma' + io)}\right) \hat{\phi}^\dagger(\varsigma') \quad (14.46)$$

Thus we can think of $\hat{\phi}^\dagger$ and $\hat{\phi}$ as raising and lowering operators and construct the Fock space as

$$|k_1, k_2 \ldots k_N\rangle = \hat{\phi}^\dagger(k_1) \hat{\phi}^\dagger(k_2) \ldots \ldots \hat{\phi}^\dagger(k_N) |0\rangle \qquad (14.47)$$

where we assume the vacuum to satisfy

$$\hat{a}(\varsigma)|0\rangle = 0 \qquad\qquad (14.48)$$

so that all the conserved quantum numbers of the vacuum would be vanishing. In terms of these new operators, the Hamiltonian of the system can be shown to take the form

$$H = \int_{-\infty}^{\infty} dk\ k^2\ \hat{\phi}^{\dagger}(k)\hat{\phi}(k) \qquad\qquad (14.49)$$

so that

$$H|k_1,k_2\ldots k_N\rangle = \left(\sum_{i=1}^{N} k_i^2\right)|k_1,k_2\ldots k_N\rangle \qquad (14.50)$$

Comparing this with Eq. (14.12), we recognize that these states are nothing other than the Bethe ansatz states up to a possible proportionality constant. This analysis brings out the connection between the Bethe ansatz and the quantum inverse scattering methods.

Yang-Baxter Equation and Quantum Groups:

In this section, we will introduce the Yang-Baxter equation and discuss very briefly the concept of quantum groups. Our discussion of quantum inverse scattering has revealed a new structure, namely, the R matrix which is defined as

$$R(\varsigma-\varsigma') = I - i\hbar r(\varsigma-\varsigma') \tag{14.52}$$

and which satisfies

$$R(\varsigma-\varsigma')\hat{T}_1(x,y;\varsigma)\hat{T}_2(x,y;\varsigma')$$

$$= \hat{T}_2(x,y;\varsigma')\hat{T}_1(x,y;\varsigma)R(\varsigma-\varsigma') \tag{14.52}$$

As we recall $\hat{T}_1$, $\hat{T}_2$ and R are defined in the tensor product space $V \otimes V$. If we now generalize these operators into the triple tensor product space $V \otimes V \otimes V$ as

$$\hat{T}_1 = \hat{T} \otimes I \otimes I$$

$$\hat{T}_2 = I \otimes \hat{T} \otimes I$$

$$\hat{T}_3 = I \otimes I \otimes \hat{T}$$

and define R_{ij}, $i,j = 1,2,3$, as the embedding of R into this space so that it acts on the ij space, then the consistency of Eq. (14.52) would imply

$$R_{12}(\varsigma-\varsigma')R_{13}(\varsigma-\varsigma'')R_{23}(\varsigma'-\varsigma'')$$

$$= R_{23}(\varsigma'-\varsigma'')R_{13}(\varsigma-\varsigma'')R_{12}(\varsigma-\varsigma') \tag{14.53}$$

This can be seen from the fact that the product $\hat{T}_1\hat{T}_2\hat{T}_3$ can be permuted to $\hat{T}_3\hat{T}_2\hat{T}_1$ in two distinct ways as

$$R_{12}(\varsigma-\varsigma')R_{13}(\varsigma-\varsigma'')R_{23}(\varsigma'-\varsigma'')$$

$$\hat{T}_1(x,y;\varsigma)\hat{T}_2(x,y;\varsigma')\hat{T}_3(x,y;\varsigma'')$$

$$= \hat{T}_3(x,y;\varsigma'')\hat{T}_2(x,y;\varsigma')\hat{T}_1(x,y;\varsigma)$$

$$R_{12}(\varsigma-\varsigma')R_{13}(\varsigma-\varsigma'')R_{23}(\varsigma'-\varsigma'') \qquad (14.54)$$

and

$$R_{23}(\varsigma'-\varsigma'')R_{13}(\varsigma-\varsigma'')R_{12}(\varsigma-\varsigma')$$

$$\hat{T}_1(x,y;\varsigma)\hat{T}_2(x,y;\varsigma')\hat{T}_3(x,y;\varsigma'')$$

$$= \hat{T}_3(x;y;\varsigma'')\hat{T}_2(x,y;\varsigma')\hat{T}_1(x,y;\varsigma)$$

$$R_{23}(\varsigma'-\varsigma'')R_{13}(\varsigma-\varsigma'')R_{12}(\varsigma-\varsigma') \qquad (14.55)$$

Comparing these two relations, we can arrive at the consistency condition of Eq. (14.53).   This relation is known as the Yang-Baxter equation after Yang and Baxter who had formulated this relation to study integrable models in statistical mechanics.   This relation is quite rich in contents and, consequently, has come up in different areas of physics and is also known as the triangle equation or the factorizability condition for the S-matrix.   We will not go into these details but let us simply note here that in the semiclassical limit, Eq. (14.53) reduces to the classical Yang-Baxter equation of Eq. (13.51), namely,

$$\left[r_{12}(\varsigma - \varsigma'), r_{13}(\varsigma) \right] + \left[r_{12}(\varsigma - \varsigma'), r_{23}(\varsigma') \right]$$
$$+ \left[r_{13}(\varsigma), r_{23}(\varsigma') \right] = 0 \qquad (14.56)$$

where we have set, for simplicity,

$$\varsigma'' = 0$$

The solutions of the classical Yang-Baxter equation lead to a classification of the classical integrable models. Since Eq. (14.55) is written completely in terms of commutators, it has a Lie algebraic structure and, consequently, its solutions can be classified using Lie algebra. On the other hand, the full Yang-Baxter equation of Eq. (14.53) is not a commutator relation and hence cannot be directly related to a Lie algebra. But let us note that it is an associative algebra which can be related to the enveloping algebra of a Lie algebra.

Given a solution of the classical Yang-Baxter equation, an interesting and open question is whether a quantum R-matrix exists, which would reduce to the given classical r-matrix. The study of this question has given rise to a very interesting mathematical structure which we describe briefly.

The Lax operator, $\hat{A}(x, \varsigma)$, acts, in principle, on two different spaces. First, there is the space of the auxiliary equation which we had chosen so far to be the two dimensional representation of SL(2). The second space is

related to the dynamical variables which themselves can
belong to some Lie algebra.   Since most continuous models
can be put on a lattice and can be identified with some spin
model, let us choose, for simplicity, the Lie algebra of the
dynamical variables also to correspond to SL(2).   Instead of
choosing the fundamental representation, we can choose any
higher representation of SL(2) for either of these spaces.
Classically, it is known that the form of the r-matrix is
insensitive to the representation.   On the other hand, it
was shown by Kulish and Reshetikhin that a solution of the
full Yang-Baxter equation exists for a higher dimensional
representation of SL(2) only if the generators of SL(2)
satisfy the following commutation relations.

$$[J_3 , J_\pm] = \pm J_\pm$$

$$[J_+ J_-] = \frac{\sinh(2\eta J_3)}{\sinh\eta}$$

$$(14.57)$$

Here η is a parameter of the theory and can in principle be
related to Planck's constant.   We note here that for the two
dimensional representation, namely, when

$$J_3 = \frac{1}{2} \sigma_3$$

the commutation relations of Eq. (14.57) reduce to the
algebra of SL(2).   Similarly, when $\eta \to 0$, the commutation
relations are precisely those of SL(2). Consequently, we
can think of Eq. (14.56) as a deformation of the enveloping
algebra of SL(2).   In the sense that quantum mechanics is a
deformation of classical mechanics, we can view Eq. (14.56)
as defining a quantum Lie algebra.   (We emphasize that Eq.

(14.56) does not define a Lie algebra. The terminology is simply borrowed from the study of quantum mechanics and is used only in the sense of describing deformations.)

This is quite interesting for it implies that the existence of a solution of the Yang-Baxter equation requires a quantization of the Lie algebraic structures. The classification of the solutions of the full Yang-Baxter equation, therefore, requires a study of quantum algebras, quantum groups and their representations. The quantum algebras have also been identified with Hopf algebras and their study is a rapidly growing area of interest.

References:

Baxter, R. J., Ann. Phys. $\underline{70}$, 193 (1972); ibid $\underline{70}$, 323 (1972).

Belavin, A. A. and V. G. Drinfel'd, Func. Anal. Appl. $\underline{17}$, 220 (1983).

Bethe, H., Z. Phys. $\underline{71}$, 205 (1931).

Bogoliubov, N. M., A. G. Izergin and V. E. Korepin, Lect. Notes Phys., Vol. 242, Ed. B. Shastry et al, Springer-Verlag, 1985.

Drinfel'd, V. G., Dokl. Akad. Nauk SSSR $\underline{283}$, 1060 (1985).

Drinfel'd, V. G. Talk at Berkeley Math. congress, 1986.

Faddeev, L. D., N.Y. Reshetikhin and L. A. Takhtajan, LOMI Preprint E-14-87, 1987.

Jimbo, M., Lett. Math. Phys. $\underline{10}$, 63 (1985); ibid $\underline{11}$, 247 (1986).

Jimbo, M., Comm. Math. Phys. $\underline{102}$, 537 (1986).

Kulish, P. P. and N. Y. Reshetikhin, J. Sov. Math. $\underline{23}$, 2435 (1981).

Kulish, P. P., N. Y. Reshetikhin and E. K. Sklyanin,
      Lett. Math. Phys. $\underline{5}$, 393 (1981).

Kulish, P. P. and E. K. Sklyanin in Lect. Notes Phys.,
      Vol. 151, Ed. J. Hietarinta et al, Springer-
      Verlag 1982.

Reshetihkin, N. Y. and L. D. Faddeev, Theo. Math. Phys.
      847 (1983).

Sklyanin, E. K., Sov. Phys. Dokl. $\underline{24}$, 107 (1979).

Sklyanin, E. K., J. Sov. Math. $\underline{19}$, 1546 (1982).

Sklyanin, E. K., Func. Anal. Appl. $\underline{16}$, 263 (1982); ibid
      $\underline{17}$, 273 (1983).

Takhtadzhyan, L. A., J. Sov. Math. $\underline{23}$, 2470 (1983).

Takhtajan, L. A. in Lec. Notes Phys., Vol. 242, Ed. B.
      Shastry et al, Springer-Verlag, 1985.

Thacker, H. B., Rev. Mod. Phys. $\underline{53}$, 253 (1981).

Thacker, H. B. in Lec. Notes Phys., Vol. 151, Ed. J.
      Hietarinta et al, Springer-Verlag, 1982.

Yang, C. N., Phys. Rev. Lett. $\underline{19}$, 1312 (1967).

APPENDIX

CALCULATION OF POISSON BRACKETS

Let us recall from Eqs. (5.47) and (5.48) that

$$\frac{\delta a(k)}{\delta u(x)} = \frac{1}{12ik} f(x,k)g(x,k)$$

and   (A.1)

$$\frac{\delta b(k)}{\delta u(x)} = - \frac{1}{12ik} f(x,k)g(x,-k)$$

We also note that if u_1 and u_2 satisfy the Schrödinger equation with eigenvalues k_1^2 and k_2^2 respectively, namely,

$$\frac{\partial^2 u_1}{\partial x^2} + \left(q + k_1^2\right) u_1 = 0$$

                                                                    (A.2)

$$\frac{\partial^2 u_2}{\partial x^2} + \left(q + k_2^2\right) u_2 = 0$$

then we can write

$$u_1 u_2 = - \frac{1}{k_1^2 - k_2^2} \frac{\partial}{\partial x} [u_1, u_2]$$ (A.3)

Consequently, given two sets of solutions of the Schrödinger equation, namely (u_1, v_1) and (u_2, v_2), we have

$$u_1 v_1 \frac{\partial(u_2 v_2)}{\partial x} - u_2 v_2 \frac{\partial(u_1 v_1)}{\partial x}$$

$$= u_1 v_1 (u_{2x} v_2 + u_2 v_{2x}) - u_2 v_2 (u_{1x} v_1 + u_1 v_{1x})$$

$$= -v_1 v_2 [u_1, u_2] - u_1 u_2 [v_1, v_2]$$

Upon using Eq. (A.3) this simplifies to

$$u_1 v_1 \frac{\partial(u_2 v_2)}{\partial x} - u_2 v_2 \frac{\partial(u_1 v_1)}{\partial x}$$

$$= \frac{1}{k_1^2 - k_2^2} \left(\frac{\partial}{\partial x} [v_1, v_2] \right) [u_1, u_2]$$

$$+ \frac{1}{k_1^2 - k_2^2} \left(\frac{\partial}{\partial x} [u_1, u_2] \right) [v_1, v_2]$$

$$= \frac{1}{k_1^2 - k_2^2} \frac{\partial}{\partial x} \left([u_1, u_2][v_1, v_2] \right) \qquad (A.4)$$

We are now ready to calculate various Poisson brackets. From the definition in Eq. (1.53), we see that

$$\{a(k), b(\ell)\}$$

$$= \frac{1}{2} \int_{-\infty}^{\infty} dx \left(\frac{\delta a(k)}{\delta u(x)} \frac{\partial}{\partial x} \frac{\delta b(\ell)}{\delta u(x)} - \frac{\partial}{\partial x} \frac{\delta a(k)}{\delta u(x)} \frac{\delta b(\ell)}{\delta u(x)} \right)$$

Upon using Eqs. (A.1) and (A.4) this can be evaluated to give

$$\{a(k) , b(\ell)\}$$

$$= \frac{1}{288k\ell} \int_{-\infty}^{\infty} dx\left[\left(f(x,k)g(x,k)\right) \frac{\partial}{\partial x} \left(f(x,\ell)g(x,-\ell)\right)\right.$$

$$\left. - \frac{\partial}{\partial x} \left(f(x,k)g(x,k)\right)\left(f(x,\ell)g(x,-\ell)\right)\right]$$

$$= \frac{1}{288k\ell(k^2-\ell^2)}$$

$$\int_{-\infty}^{\infty} dx \frac{\partial}{\partial x} \left([f(x,k),f(x,\ell)][g(x,k),g(x,-\ell)]\right)$$

$$= \frac{1}{288k\ell(k^2-\ell^2)} \left([f(x,k),f(x,\ell)]\right.$$

$$\left. [g(x,k),g(x,-\ell)]\right)\Big|_{-\infty}^{\infty} \quad (A.5)$$

We can evaluate this explicitly using Eq. (5.8) and the asymptotic forms of the Jost functions. Thus, for example,

$$\lim_{x\to\infty} \frac{1}{288k\ell(k^2-\ell^2)} [f(x,k),f(x,\ell)][g(x,k),g(x,-\ell)]$$

$$= \lim_{x\to\infty} \frac{1}{288k\ell(k^2-\ell^2)} \left[a(k)e^{-ikx}-b(-k)e^{ikx},\right.$$

$$\left. a(-\ell)e^{i\ell x}-b(\ell)e^{i\ell x}\right]$$

$$= \lim_{x \to \infty} \frac{i(k-\ell)e^{i(k+\ell)x}}{288k\ell(k^2-\ell^2)} \left[-i(k+\ell)a(k)a(-\ell)e^{-i(k-\ell)x} \right.$$

$$+ i(k-\ell)a(k)b(\ell)e^{-i(k+\ell)x}$$

$$- i(k-\ell)a(k)b(\ell)e^{-i(k+\ell)x}$$

$$\left. + i(k+\ell)b(-k)b(\ell)e^{i(k-\ell)x} \right]$$

$$= \lim_{x \to \infty} \frac{i(k-\ell)}{288k\ell(k^2-\ell^2)} \left[-i(k+\ell)a(k)a(-\ell)e^{2i\ell x} \right.$$

$$+ i(k-\ell)a(k)b(\ell)$$

$$-i(k-\ell)a(-\ell)b(-k)e^{2i(k+\ell)x}$$

$$\left. + i(k+\ell)b(-k)b(\ell)e^{2ikx} \right]$$

If we further use the relation,

$$\lim_{x \to \infty} \frac{e^{ikx}}{k} = i\pi\delta(k)$$

then the above expression becomes, for $k, \ell > 0$,

$$\lim_{x \to \infty} \frac{1}{288k\ell(k^2-\ell^2)} [f(x,k),f(x,\ell)][g(x,k),g(x,-\ell)]$$

$$= -\frac{(k-\ell)}{288k\ell(k+\ell)} a(k)b(\ell) \qquad k,\ell > 0 \qquad (A.6)$$

Similarly, we can show that for $k,\ell > 0$

$$\underset{x\to-\infty}{\text{Lim}} \frac{1}{288k\ell(k^2-\ell^2)} [f(x,k),f(x,\ell)][g(x,k),g(x,-\ell)]$$

$$= \frac{(k+\ell)}{288k\ell(k-\ell)} a(k)b(\ell) - \frac{i\pi}{144k} a(k)b(k)\delta(k-\ell) \quad (A.7)$$

It then follows from Eqs. (A.6) and (A.7) that for $k,\ell > 0$

$$\{a(k) , b(\ell)\} = - \frac{1}{144} \frac{k^2+\ell^2}{k\ell(k^2-\ell^2)} a(k)b(\ell)$$

$$+ \frac{i\pi}{144k} a(k)b(k)\delta(k-\ell) \quad (A.8)$$

Similarly, we can also show that for $k,\ell > 0$

$$\{a(k) , b(-\ell)\} = \frac{1}{144} \frac{k^2+\ell^2}{k\ell(k^2-\ell^2)} a(k)b(-\ell)$$

$$- \frac{i\pi}{144k} a(k)b(-k)\delta(k-\ell) \quad (A.9)$$

Let us next note that for $k,\ell > 0$

$$\left\{\log a(k) , \log \frac{b(\ell)}{b(-\ell)}\right\} = \{\log a(k) , 2i \arg b(\ell)\}$$

$$= 2i\{\log|a(k)| , \arg b(\ell)\}$$

$$- 2\{\arg a(k) , \arg b(\ell)\} \quad (A.10)$$

On the other hand, using Eqs. (A.8) and (A.9) we get

$$\left\{ \log a(k) \ , \ \log \frac{b(\ell)}{b(-\ell)} \right\}$$

$$= \{\log a(k) \ , \ \log b(\ell)\} - \{\log a(k) \ , \ \log b(-\ell)\}$$

$$= \frac{1}{a(k)b(\ell)} \{a(k),b(\ell)\} - \frac{1}{a(k)b(-\ell)} \{a(k),b(-\ell)\}$$

$$= -\frac{1}{144} \frac{k^2+\ell^2}{k\ell(k^2-\ell^2)} + \frac{i\pi}{144k} \delta(k-\ell)$$

$$- \frac{1}{144} \frac{k^2+\ell^2}{k\ell(k^2-\ell^2)} + \frac{i\pi}{144k} \delta(k-\ell)$$

or, $\left\{ \log a(k) \ , \ \log \frac{b(\ell)}{b(-\ell)} \right\}$

$$= \frac{i\pi}{72k} \delta(k-\ell) - \frac{1}{72} \frac{k^2+\ell^2}{k\ell(k^2-\ell^2)} \tag{A.11}$$

Comparing with Eq. (A.10), we conclude that for $k,\ell > 0$

$$\{\log|a(k)| \ , \ \arg b(\ell)\} = \frac{\pi}{144k} \delta(k-\ell) \tag{A.12}$$

$$\{\arg a(k) \ , \ \arg b(\ell)\} = \frac{1}{144} \frac{k^2+\ell^2}{k\ell(k^2-\ell^2)} \tag{A.13}$$

We can similarly calculate other Poisson brackets and show that for $k,\ell > 0$

$$\{a(k) \ , \ a(\ell)\} = 0 = \{b(k) \ , \ b(\ell)\} \tag{A.14}$$

$$\{\arg b(k) \ , \ \arg b(\ell)\} = 0 \tag{A.15}$$

$$\{\log|a(k)| \ , \ \log|a(\ell)|\} = 0 \qquad\qquad (A.16)$$

Thus we see that for $k, \ell > 0$, if we choose the variables

$$Q(k) = \arg b(k)$$
$$\qquad\qquad (A.17)$$
$$P(k) = -\frac{144k}{\pi} \log |a(k)|$$

then they would satisfy

$$\{Q(k) \ , \ Q(\ell)\} = 0 = \{P(k) \ , \ P(\ell)\}$$
$$\qquad\qquad (A.18)$$
$$\{Q(k) \ , \ P(\ell)\} = \delta(k-\ell)$$

In other words, the variables defined in Eq. (A.17) constitute a canonical set. Incidentally, we can also calculate in a straightforward manner and show that for $k, \ell > 0$

$$\{\log a(k) \ , \ \log a(\ell)\} = 0 \qquad\qquad (A.19)$$

It is clear from this that if $\log a(k)$ generates conserved quantities, then these conserved quantities would be in involution.

We can determine the canonical set of variables from the discrete spectrum in the following way. Let us note from Eq. (7.29) (The derivation is straightforward from the Schrödinger equation.) that for the nth bound state we have

$$\frac{\delta\lambda_n}{\delta u(x)} = -\frac{1}{6} c_n f^2(x, i\kappa_n)$$

where   (A.20)

$$c_n = \left(\int_{-\infty}^{\infty} dx \; f^2(x, i\kappa_n)\right)^{-1}$$

Recalling that

$$\lambda_n = -\kappa_n^2$$

we conclude that

$$\frac{\delta\kappa_n^2}{\delta u(x)} = \frac{1}{6} c_n f^2(x, i\kappa_n) \qquad\qquad (A.21)$$

Similarly, recalling that b_n is the analytic continuation of $b(k)$ to $k = i\kappa_n$, we see from Eq. (A.1) that

$$\frac{\delta b_n}{\delta u(x)} = \frac{1}{12\kappa_n} f(x, i\kappa_n) g(x, -i\kappa_n) \qquad (A.22)$$

The Poisson brackets are now easy to calculate. For example, for $n \neq m$, we have

$$\{b_n, \kappa_m^2\} =$$

$$= \frac{1}{2} \int_{-\infty}^{\infty} dx \left(\frac{\delta b_n}{\delta u(x)} \frac{\partial}{\partial x} \frac{\delta \kappa_m^2}{\delta u(x)} - \frac{\partial}{\partial x} \frac{\delta b_n}{\delta u(x)} \frac{\delta \kappa_m^2}{\delta u(x)} \right)$$

$$= \frac{c_m}{24\kappa_n} \int_{-\infty}^{\infty} dx \left[\left(f(x,i\kappa_n)g(x,-i\kappa_n) \right) \frac{\partial}{\partial x} \left(f^2(x,i\kappa_m) \right) \right.$$

$$\left. - \frac{\partial}{\partial x} \left(f(x,i\kappa_n)g(x,-i\kappa_n) \right) f^2(x,i\kappa_n) \right]$$

Using Eq. (A.4) this can be written as

$$\{ b_n , \kappa_m^2 \}$$

$$= \frac{c_m}{144\kappa_n(\kappa_n^2-\kappa_m^2)} \int_{-\infty}^{\infty} dx \frac{\partial}{\partial x} \left([f(x,i\kappa_n),f(x,i\kappa_m)] \right.$$

$$\left. [g(x,-\kappa_n),f(x,i\kappa_m)] \right)$$

$$= \frac{c_m}{144\kappa_n(\kappa_n^2-\kappa_m^2)} \left([f(x,i\kappa_n),f(x,i\kappa_m)] \right.$$

$$\left. [g(x,-i\kappa_n),g(x,i\kappa_m)] \right) \Big|_{-\infty}^{\infty} \quad \text{(A.23)}$$

Since the bound state wave functions vanish asymptotically, we conclude that for $n \neq m$

$$\{ b_n , \kappa_m^2 \} = 0 \qquad\qquad\qquad \text{(A.24)}$$

On the other hand,

$$\{b_n , \kappa_n^2\}$$

$$= \frac{1}{2} \int_{-\infty}^{\infty} dx \left(\frac{\delta b_n}{\delta u(x)} \frac{\partial}{\partial x} \frac{\delta \kappa_n^2}{\delta u(x)} - \frac{\partial}{\partial x} \frac{\delta b_n}{\delta u(x)} \frac{\delta \kappa_n^2}{\delta u(x)} \right)$$

$$= \frac{1}{2} \int_{-\infty}^{\infty} dx \left[\frac{\delta \kappa_n^2}{\delta u(x)} , \frac{\delta b_n}{\delta u(x)} \right]$$

$$= \frac{c_n}{144\kappa_n} \int_{-\infty}^{\infty} dx \left[f^2(x,i\kappa_n) , f(x,i\kappa_n)g(x,-i\kappa_n) \right]$$

$$= \frac{c_n}{144\kappa_n} \int_{-\infty}^{\infty} dx \; f^2(x,i\kappa_n) \left[f(x,i\kappa_n) , g(x,-i\kappa_n) \right]$$

From the relation in Eq. (5.9) we see that the above expression can be simplified to

$$\{b_n , \kappa_n^2\} = \frac{c_n b_n}{72} \int_{-\infty}^{\infty} dx \; f^2(x,i\kappa_n)$$

On the other hand, from the definition of c_n in Eq. (A.20), we readily see that we can write

$$\{b_n , \kappa_n^2\} = \frac{b_n}{72} \tag{A.25}$$

Thus we can write

$$\{b_n , \kappa_m^2\} = \frac{b_n}{72} \delta_{n,m} \tag{A.26}$$

It is easy to show that (we have already shown the second relation in Eq. (7.40).)

$$\left\{ b_n \ , \ b_m \right\} = 0 = \left\{ \kappa_n^2 \ , \ \kappa_m^2 \right\} \tag{A.27}$$

It is clear, therefore, that if we choose

$$q_n = \frac{1}{2} \log |b_n|$$

$$\tag{A.28}$$

$$p_n = 144 \ \kappa_n^2$$

then they would satisfy

$$\{q_n \ , \ q_m\} = 0 = \{p_n \ , \ p_m\}$$

$$\tag{A.29}$$

$$\{q_n \ , \ p_m\} = \delta_{n,m}$$

In other words, this constitutes a canonical set also. Together $(Q(k), q_n, P(k), p_n)$ constitute the action angle variables of the system. It is a complete set since the scattering data forms a complete set.

References:

Faddeev, L. D. and L. A. Takhtajan, Lett. Math. Phys. 10, 183 (1985).

Novikov, S., S. V. Manakov, L. P. Pitaevskii and V. E. Zakharov, Theory of Solitons, Consultants Bureau, 1984.

Zakharov, V. E. and S. V. Manakov, Theo. Math. Phys. 19, 551 (1974).

INDEX

Abelian current algebra
  17
Action-angle variables
  10, 105, 292
AKNS method, 253
Alternating step function
  19, 208, 308
Auxiliary linear problem
  310
Bäcklund transformation
  162
    for KdV equation, 174
    for Liouville equation
      164
    for Sine-Gordon
    equation, 168
Bethe ansatz, 304
Bianchi identity, 6, 198
Bound states, 43, 80
    normalization of
      84, 88, 101
    of soliton potential
      43
Bose gas, 306
Canonical Poisson brackets
  5, 246
Cartan-Maurer equation
  155
Cartan subalgebra, 234
Cartan-Weyl basis, 234
Chevalley basis, 239
Classical Yang-Baxter
equation, 292, 323
Conserved densities
  56, 113

Conserved quantities
  8, 46, 109, 203, 219,
  245, 300
Darboux's theorem, 196
Delta function interaction
  306
Dispersion relation, 33
Dual Poisson bracket
  20, 196, 202, 215
Factorizability condition
  324
Frechet derivative
  141, 147
Functional derivative
  15, 141
Functional recursion
relation, 67, 158
Fundamental Poisson
bracket, 16, 49, 288, 291
Galilean transformation
  54
Gel'fand-Levitan equation
  85, 89
Generalized coordinates
  3, 14, 214
Geometry
    of phase space, 189
    Riemannian, 6
    Symplectic, 6, 189
Geometrical approach
    to integrable models
      189
    to KdV, 207
Group structure of Toda
equation, 240

Hamiltonian system, 2, 6
    integrability of, 7
    KdV as, 14
    NSE as, 265
Hamiltonian vector field
 195
Higher order equation
 71, 163
Hopf algebra, 325
Integrability
    of Hamiltonian system
    7
    of KdV equation
    65, 70
    of MKdV equation, 70
    of NSE, 300
    of Toda lattice
    225, 247
    quantum, 303
Inverse scattering
 92, 266, 309
Involution
 8, 70, 150, 205, 225,
 251, 298, 333
Isospectral, 40, 123
Jacobi identity
 6, 193, 235, 291
Jost functions, 93 270
Korteweg-de Vries (KdV)
equation, 11
    Bäcklund transfor-
    mation of, 174
    conserved densities
    56, 61

    integrability of, 65
    Lax pair for, 125
    Lenard's derivation of
    135
    modified, 51
    multi-soliton solution
    of, 180
    Poisson bracket of
    16, 20
    symmetries of, 12
KdV solutions
    properties of, 24
    uniqueness of, 24
Lagrangian
    for KdV, 21
    for NSE, 265
    for the Toda lattice
    215, 216
Lax pair
 121, 226, 244, 255, 258
Lie algebra, 153, 232, 324
Lie group, 153, 233
    dimension of, 233
    rank of, 233
    structure constants of
    233
Light cone variables
 163, 263, 264
Liouville equation, 163
Liouville's theorem, 6
Matrix Jost solutions, 284
Miura transformation
 51, 174
MKdV equation, 51

Monodromy matrix, 285

Multi-soliton solutions
162, 187

N-body Schrödinger
Hamiltonian, 306

Nijenhuis torsion tensor
204, 232

Nonlinear Schrödinger
equation, 262

Permutation matrix
289, 293

Poisson brackets
dual
20, 196, 202, 215
of KdV, 16, 20
of NSE, 266
of Toda lattice, 216

Positive roots, 236

Quantum algebra, 325

Quantum group, 325

Quantum integrability, 303

Quantum inverse scattering
309

r matrix, 248, 291, 323

R matrix, 310, 321

Reflectionless potential
44

Riccati relation, 52, 117

Root vectors, 235

Scaling, 11

Scattering data
102, 115, 275, 285

Simple roots, 237, 248

Sine-Gordon equation
167, 263

Sinh-Gordon equation, 264

SL(2), 325

SL(2,R), 153

Solitons
10, 30, 34, 44, 162, 180

Spectral parameter
119, 140, 257

SU(2), 289

SU(N), 232

Symplectic diffeomorphism
195

Symplectic form, 215

Symplectic geometry
6, 189

Symplectic manifold
6, 189

Symplectic metric, 7, 194

Symplectic structure, 193

Theorem of permutability
171, 182

Toda equation, 213

Toda lattice, 213
dual Poisson bracket
structure of, 215
group structure of
240
Hamiltonian of
215, 216
integrability of
225, 247
Lagrangian of
215, 216
Lax equation for
225, 245

    Lax pair for, 226, 244
    Nijenhuis tensor for,
     221
Transition matrix
 275, 281, 284
    quantum, 311
Triangle equation, 324
Uniqueness
    of conserved
    quantities, 64
    of solutions, 24
Wronskian, 94, 269
Yang-Baxter equation
 292, 322
Zakharov-Shabat
formulation, 253, 278
Zero curvature
    condition, 153
    method, 278